USER

INFOTECHNODEMO

MEDIAWORKBOOK

BY
PETER LUNENFELD

VISUALS BY
MIEKE GERRITZEN

PUBLISHED BY
THE MIT PRESS
CAMBRIDGE, MASSACHUSETTS
LONDON, ENGLAND

AND

ALL MEDIA FOUNDATION
AMSTERDAM

COGITOERGOSUMOCOGITOERG
INFOTECHNODEMOINFOTECHN
NEURONANOPORNONEURONAN
CRYPTOMICROMACROCRYPTO
MONDOMYSTOUFOMONDOMYS
NEOGEOECONEOGEOECONEOG
PROTOPROMOSOLOPROTOPRO
CRYOHYPNOHEROCRYOHYPNO
METRORETROPSYCHOMETROR
GLOBOEUROLINGOGLOBOEURO
KLEPTOPINKOBUNCOKLEPTOP
LOCOPYROSTUDIOLOCOPYROS
PALEOROBODISCOPALEOROBO
NARCOSACROTHEONARCOSAC
GENOTYPOLOGOGENOTYPOLO
EXOMYTHOMEMOEXOMYTHO
ASTROMOJORADIOASTROMOJ
ETHNOBIOEGOETHNOBIOEGOE
AEROAUTONATOAEROAUTONA
XENOZEROSTEGOXENOZEROST
ANDROGYNOVIDEOANDROGYN
SCHIZOWINONYMPHOSCHIZOW
COGITOERGOSUMOCOGITOERG
INFOTECHNODEMOINFOTECHN

USER

BY PETER LUNENFELD

VISUALS: MIEKE GERRITZEN

Library of Congress Cataloging-in-Publication Data

Lunenfeld, Peter
User : InfoTechnoDemo / Peter Lunenfeld ;
visuals by Mieke Gerritzen
ISBN 0-262-62198-3 (pbk : alk. paper)
1. Computers and civilization. 2. Digital Media.
3. Technology, Social aspects. I. Title
QA76.9.C66.L8623 2005
303.483 dc22

OGITOERGOSUMOCOGITOERGO
NFOTECHNODEMOINFOTECHNO
EURONANOPORNONEURONANO
RYPTOMICROMACROCRYPTOM
IONDOMYSTOUFOMONDOMYST
EOGEOECONEOGEOECONEOGEO
ROTOPROMOSOLOPROTOPROM
RYOHYPNOHEROCRYOHYPNOH
ETRORETROPSYCHOMETRORE
LOBOEUROLINGOGLOBOEUROL
LEPTOPINKOBUNCOKLEPTOPIN
OCOPYROSTUDIOLOCOPYROST
ALEOROBODISCOPALEOROBOD
ARCOSACROTHEONARCOSACR
ENOTYPOLOGOGENOTYPOLOGO
XOMYTHOMEMOEXOMYTHOM
STROMOJORADIOASTROMOJO
THNOBIOEGOETHNOBIOEGOETH
EROAUTONATOAEROAUTONATO
ENOZEROSTEGOXENOZEROSTE
NDROGYNOVIDEOANDROGYNOV
HIZOWINONYMPHOSCHIZOWIN
OGITOERGOSUMOCOGITOERGO
IFOTECHNODEMOINFOTECHNO

INTRODUCTION: UTILITIES, NOT MANIFESTOS

looks at art and video-games, book design and techno-masturbation, comix and life extension diets.

readers will have to determine for themselves if this range is symptomatic of pluralism or promiscuity. The point is to make visible the patterns and repetitions of our moment that link nanotechnology to electronic music, artist/archivist Harry Smith to architect/superstar Rem Koolhaas, and Pontiacs to open source software. My writings reflect an obsession with doing theory and criticism in real time, which is akin to holding mercury in my fingers. But, like mercury, some of this stuff is toxic, and I'd be as happy never thinking about it again. I'm not saying that paintings = industrial design = *The Matrix* = Web porn, just that all of these and more exist in the same lives – mine, hers, theirs, perhaps yours.

USER's essays batter themselves against that overwhelming diversity which for lack of a better name we call the present. The present offers a sense of exhilarating unsustainability with genuine intellectual and aesthetic challenge, tempered by the surety that it isn't going to last. Part of the excitement is that the old categories don't hold; there's no sense in talking about coherent, oppositional avant-gardes anymore. There are still cries of rage and acts of violence, but not much in the way of compelling manifestos. I'm interested in utilities anyway, not manifestos.

A utility adds functionality to a software system, and can be built upon in turn by further utilities later on down the road. Utilities are evolutionary rather than revolutionary, unlikely to call for the destruction of a system in order to save it. These essays then, are translator utilities, bridgin divides between the art world and the design establishment, between journalism and the seminar room. They were written over a five year period for *artext*, the Los Angeles-based international magazine. A regular column offers the writer a chance t fashion a dialogue over time with readers, to have the essays speak to and through each other. This form is bound up in its own constraints. One of the toughest is th question of whether critics can engage in real time with popular culture without bein overwhelmed by its banality. For the utilities in *USER* to function as criticism, they should be read with the knowledge that while produced within the synergized environments they describe, they point toward the future's very different present.

Thanks are due to Joan Shigekawa and the Rockefeller Foundation for their support of the Mediawork project; to Paul Foss, the publisher of *artext*, for his generosity, insight and support; and to Doug Sery at the MIT Press for his courageous commitment to the marriage of writing and design. The transformation of "User," the column, into *USER*, the Mediawork book, began when Geert Lovink brought me together with Mieke Gerritzen, who understands so well how the right fusion of words and images becomes as potent as a smart bomb. Finally, none of this could have happened without Susan Kandel, who, as the editor of *artext*, had commissioned these essays, but who, as my wife, had to put up with me when I was actually writing and revising them.

PERMANENT PRESENT

PERMANENT PRESENT

PERMANENT PRESENT

USER *PERMANENT PRESEN*

20202020202020202020202020202
202222202020202222020002020202
000022220202020202020002022202
02020202020202002222020200022
202000222202020202220202022020
202002002002000020202202022202
202020200202002020020202020202
020202002020202020202020202020
20202020202020202222202020202

HAVING SPED PAST
THE MANUFACTURED MILESTONE
OF THE MILLENNIUM

222020200022222020002222020202
220202022020202020020020020000
202202022202002020202002020020
020202020202020202020020202020
202020202020202020202020202020
222020202022220200020202020000
222202020202020200020222020202
02020202020022220202000222220
0022220202020222020202220202020
0200200200002020222020222020020
202002020020200202020202020202
020020202020202020202020202020
202020202020222220202020222202
020202020000002222020202020202

OUR VISUAL CULTURE

REMAINS TRAPPED

IN A RELENTLESS PRESENT

IDLY CIRCLING ITSELF AS

IF WAITING

FOR INSPIRATION

IT DOESN'T EXPECT TO COME

For good or ill,
the high-modern period
offered a succession of
startling visions
of what was to come,

indeed, the future of the future seemed assured precisely because the scenarios themselves were so heterogeneous.

USER PERMANENT PRESEN

Nikolai foregger's "machine dances," concocted in the heat of SOVIET REVOLUTIONARY IDEALISM, and LE CORBUSIER'S slick materials fetish of the 1930's both extolled the glories of th industrial, though to radically different effects:

FLASH GORDON'S *pulp imaginary* on Planet Mongo battled it out with corporatized moonscape in STANLEY KUBRICK'S 2001:

Futurama at the 1939 World's Fai presaged, but remained distinct from, the google diners and gas stations of Southern California's atomic age:

★★★★★★★★★★★★★★★

HUGH HEFNER'S space-age Playboy bachelorpads emerged during the same period as their virtual antitheses, the ecotopian fantasies of the hippies and greens.

IF, FROM THE TURN OF THE CENTURY THROUGH THE 1960'S, FUTURES WERE MAPPED OUT, PAINTED, PRINTED, FILMED, ANIMATED, AND EVEN OCCASIONALLY BUILT, TODAY THE FEVER FOR WHAT IS YET TO BE HAS PERCEPTIBLY COOLED. THERE ARE REASONS FOR THIS, OF COURSE: A DISAPPOINTMENT THAT THE PRESENT HAS NOT LIVED UP TO THE PAST'S HOPES FOR IT; PERHAPS AN EXASPERATION WITH THE 20TH CENTURY'S PROLIFERATION OF OPTIONS AND BRUTAL PACE OF CHANGE. THAT SAID, I LAY BLAME FOR OUR 21ST CENTURY INABILITY TO IMAGINE ANYTHING BEYOND THE MOMENT ON TWO, ALMOST PERFECT VISUAL SYSTEMS, BOTH MOVING INTO THEIR THIRD DECADE.

THE FIRST IS A MOVIE;

THE SECOND, AN INTERFACE.

Consider the problem of Ridley Scott's *Blade Runner*, released in 1982. With its skyscraper corridors, glaring red neon, and mix of 1940's padded shoulders and Japanese high-tech gloss, *Blade Runner's* aesthetic – described by Syd Mead, the visual futurist for the film, as "a maze of mechanical detail overlaid onto barely recognizable architecture producing an encrusted combination of style which we...labeled 'RETRO-DECO'" – was perfectly postmodern. As such, it has inoculated every cinematic depiction of the future to emerge since. When Robert Longo bought the rights to "Johnny Mnemonic," William Gibson's seminal 1981 cyberpunk short story, no one was expecting mastery, but Longo's 1995 film was a textbook case of *Blade Runner* blindness: the only thing it did was make you wish you were watching the director's cut of Scott's opus. Audiences rightly marveled at *The Matrix* (though less so its sequels *The Matrix Reloaded* and *The Matrix Revolutions*) because of the ways the Wachowski Brothers combined Hong Kong balletics with the meta-human capacities that comic book readers have always had to imagine for themselves.

But the directors and their production team were incapable, through three films, with two worlds at their disposal, to make either one look like more than mediocre rip-offs of RETRO-DECO. Just as the *Ring* cycle virtually destroyed grand opera after Wagner, *Blade Runner* is *gesamt*, so complete in its conception, execution and integration, that other filmmakers have either refused to compete or failed miserably in the attempt.

Paradoxically, *Blade Runner* truncated the possibilities for cinematic visions of the future at the very moment that science fiction became the dominant genre in Hollywood. Ever since George Lucas proved unenthusiastic industry executives wrong with the original *Star Wars* trilogy, SF has dominated the international box office, with massive franchises like the *Terminator*, *Jurassic Park*, *Alien*, *Predator*, and *Men in Black* films. Yet for all its commercial success, the SF boom has been anti-visionary: the future is Now combined with Then, with new weapons and ever freakier aliens.

During the period of SF's ascendancy, much was made of the importance of digital technologies to the genre's success. But what these tools made possible was a digital realism so total that rather than expand the palette of choices, it paradoxically shut them down. (And what of digital surrealism? Its possibilities have yet to be imagined.) Which leads us to sticking point number two: the computer interface we all know and love/hate. While there are still a few command line systems left in use (think back to the ubiquitous C:// prompt),

THE MARKET IS DOMINATED BY ONE PARTICULAR FLAVOR OF GRAPHICAL USER INTERFACE, OR GUI, AS IT'S CALLED

THIS IS THE DESKTOP METAPHOR OF

NESTED FILES

ICONS

TRASH CANS

CASCADING WINDOWS

Masterminded in the '60s and perfected for the personal computer at Xerox PARC in the '70s, it has become the definitive human/ computer interface since Apple introduced the Macintosh operating system in '84 and Microsoft rode it to hegemony with Windows. For all its success, this interface model, like *Blade Runner*, has become an impediment to thinking beyond the present.

While there are regular software upgrades that incrementally increase design efficiency, nothing radically different has penetrated the computer market, nothing that screams:

THIS IS THE
FUTURE
PAY
ATTENTION

If *Blade Runner* stymied innovation by inspiring aesthetic terror in makers, the very notion of transcending the Windows interface inspires horror in users. Techno-anxiety is the justifiable fear of novelty, and of integrating the new into one's life, especially when that new involves what is for most people the technological unknown. The preference for the comfortable if back-assed wins out over elegance of interaction or metaphorical brilliance every time. This phobic avoidance of new tools, then, has naturalized one type of interface in the very infancy of electronic information design.

SER
PERMANENT PRESENT

THESE DAYS, EVEN THE PEOPLE WE PAY TO CONSTRUCT VISIONS OF THE FUTURE HAVE ABDICATED THEIR RESPONSIBILITIES.

Not even the plucky Disney imagineers–who claim to forge architecture, technology, and entertainment (the Alberichs of scripted spaces, to return to Wagner)–have lost the will.

When they built Euro-Disney in the '80s, **THEIR TOMORROWLAND WAS CONSTRUCTED** not from the stuff of their own researches and fantasies, but instead around the future imagined by nineteenth-century visionaries like Jules Verne.

For its ***NEXT VENTURE***, Disney ought recruit Matthew Barney, whose *Cremaster* films (1994-2002) are the only thing I've seen in a decade **to offer a credible vision of otherness, if not the future per se.**

Cremaster 4, the first to be shot, remains emblematic.

Filmed on the Isle of Man, it features a local zoological anomaly with one horn pointing up and the other down, known as the Loughton ram; free-floating testicles; androgynous, body-building fairies in high heels; and Barney himself as the anti-heroic "Loughton Candidate," in a white suit, spats, and prosthetic facial makeup.

WITH BARNEY, HOWEVER, WE QUICKLY SEE THE FOLLY OF STRAIGHT EXEGESIS.

THE *CREMASTER* FILMS ARE *PHANTASMAGORICALLY OVER-CODED,* *MAKING REFERENCE TO EVERYONE FROM HARRY HOUDINI TO GARY GILMORE, SIGN SYSTEMS FROM FREEMASONRY TO MOTORCYCLE RACING, EVERYTHING FROM VASELINE TO A HYBRID CHEETAH-WOMAN. BACKGROUND FACTS AND INFORMATION CAN INDEED BE USEFUL ROUTES INTO THE FILMS.*

But to think that this art system is fully legible to anyone (up to and including their maker) seems a critical error.

These are dynamic sound image matrices which offer a wide series of pleasures and a depthless mystery. With *Cremaster*, Barney created his own weird cosmogony – neither science-fiction nor realism, one oozing with meaning and dripping with metaphor.

It offers a way out of our permanent present because it enfolds the biological and the genetic, which are to the twenty-first century what the mechanical and the cybernetic were to the last one.

USER
SOLITUDE
ENHANCEMENT
MACHINES
USER

History is spunk
OR, AT LEAST THE HISTORY OF MEDIA IS.
The technics of what a friend once delicately referred to as
"solitude enhancement"
DRIVE THE SUCCESSES AND FAILURES OF VAST ENTERPRISES.
$ $ $ $
ONAN'S HEIRS ADOPT TECHNOLOGIES EARLY,
PAYING A PREMIUM FOR SELF-PLEASURE,
sustaining media until
THEY BROADEN
THEIR APPEAL
AND

OLITUDE ENHANCEMENT MACHINES

DROP THEIR

PRICE POINT

Home video was erected on just such a sticky base (writing about bachelors' machines tends to bring out sniggers in us all). In the first five years or so after the introduction of home video machines, a full eighty per cent of those who owned a player owned one or more pornographic tapes, a percentage that was not coincidental.

NOT JUST THE MEDIUM

BUT ITS DOMINANT STANDARD DEPENDED ON THE BACHELORS' LOYALTIES. VHS CRUSHED BETAMAX ALMOST A QUARTER OF A CENTURY AGO, BUT SONY'S BETA IS BEMOANED BY THOSE WHO LOVE THE MYTH OF THE BEAUTIFUL LOSER. HOW COULD THE PUBLIC HAVE CHOSEN VHS'S LOWER RESOLUTION, WORSE SOUND, AND FLIMSIER TAPES?

There is a standard economic answer:

SONY maintained sole rights to its elegant technology, while VHS was licensed to any and all comers, driving down costs and increasing availability (if this sounds like Apple's losing strategy for the Macintosh operating system versus the hegemony of Microsoft Windows, that's because it is). There is a further reason for the victory of VHS's crappy technology, though. SONY, wary of boycotts, made it difficult for North Hollywood's pornographers to buy tape duplication equipment in bulk, while the diversified producers of VHS equipment were more than happy to fill the demand.

The resulting VHS pornucopia was the sexually transmitted disease that killed off Beta. Promoters of DVDs, SONY included, learned their lesson, and there has never been a dearth of sex on disc.

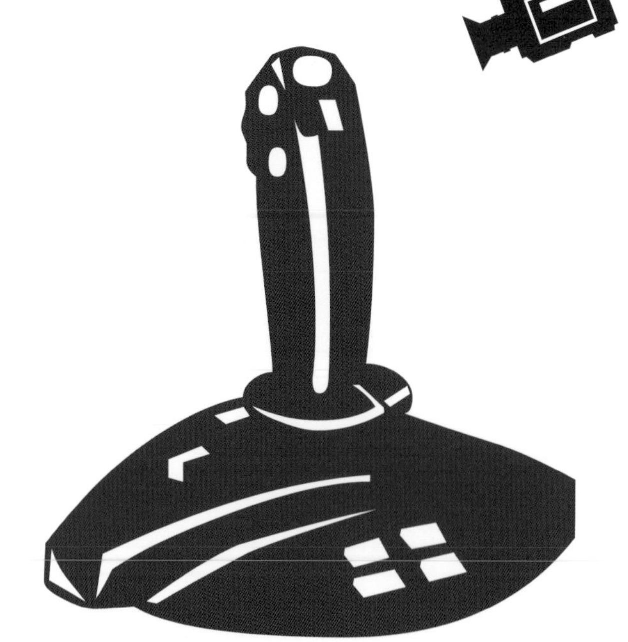

This pornucopia is not limited to video technologies, of course. Sex media have been vital to the imagescape since Daguerre. In 1870, visitors to Paris craved photographic French postcards; in 1887, Eadweard Muybridge made his "scientific" motion studies of naked women; in 1927, Elks Club smokers featured stag films; in 1972, that kind of hard core moved into movie theaters with *Deep Throat*; by 1977, cable television was offering "explicit" materials both by subscription and as pay-per-view. While mainstream Hollywood continues to explore the possibilities of reducing theatrical distribution costs by eliminating celluloid, what few adult houses remain long ago switched over to large-scale video projection. The high expense, low resolution, and bandwidth limitations of virtual reality systems prevented a profusion of sexualized VR (though the fantasy of a full-scale, interactive, immersive and even tactile tele-dildonics intoxicated press and public alike), but bulletin board systems (BBS), proprietary online services like America On Line (AOL), and the World Wide Web (WWW) all capitalized nicely on the desires of the one-handed typist. Though one never finds mention of it in the corporate literature, AOL owed much of its early success to middle-aged men and teenaged boys spending endless (billable) hours in sex chat rooms.

Psychologists, sexologists and sociologists are just beginning to investigate the effect on nominatively normative heterosexual males of pretending to be each other's hot-'n'-horny, busty female love slaves (a homosocial drag show moving into its second decade of complete disavowal).

THE BEST THING ABOUT QUOTING

ANDY WARHOL

is that you can never be sure if he actually said or wrote what is attributed to him.

SO, WHEN I SAY THAT SAINT ANDY DEFINED THE AGE AS ONE IN WHICH WE ARE

"BORED AND HYPER"

AT THE SAME TIME, WHO CAN ARGUE?

Posterboys for the bored and hyper, the bachelors anxiously wait to download their umpteenth poorly scanned jizz pic, their bandwidth-hogging video clips.

In thrall to zipless algorithms, they graze sex media, ever restive, never satisfied.

SOLITUDE ENHANCEMENT MACHINE

The Web's creation of micro-communities is finally bringing the bachelors out of their closets (or dens, in the case of the family men). Witness the <alt.sex.masturbation> discussion group, which should probably be under the <rec.arts> subheading, and the obsessive coverage of sex media, with the major adult movie reviews archive cribbing for its motto Pauline Kael's remark that the critic's independence is everything, while all else is marketing. On the other hand, while sex media content undergoes continued refinement (or degradation, if that's the way you prefer to see it), delivery systems stagnate. Where are the industrial designers who will honor modernism's call to functionality? We have ergonomic work stations, but no wired masturbation couches. Women and gay men tend to be far more open about technologies like vibrators and sex toys than straight men, who often laugh off the purchase, rental, and downloading of explicit imagery as a joke or momentary diversion. Making the commitment to dedicated auto-erotic systems would render such demurrals ludicrous.

For more than a decade, New York has been criminalizing the city's formerly thriving sex industry, but it was in that city's Sandra Gering Gallery that I saw a sliver of the future. Jordan Crandall's *DRIVE* is nominally about the transformation of the cinematic into the visual database, with 16 and Super 8 film transferred to digital video, assembled on a nonlinear editing system, run through military and motion tracking software, mastered to Digital Video Disc (DVD), and then deployed in three different ways: as static images arranged in light boxes, as quasi-filmic video projections, and as portable, personal image immersions. This last was accomplished with the loan from SONY of their portable DVD players and the Glasstron, a head-mounted display system complete with earphones jacked into the DVD player. Crandall had a cozy relationship with SONY, which sponsored his installation at Documenta X, and I'm not sure what their executives made of his show, though it shouldn't have offended them in any way. I'm more interested in whether they were aware of how close they'd come to marketing a component of a universal masturbation device.

SOLITUDE ENHANCEMENT MACHINE

Connected to a portable PC with wireless capacity and a built-in DVD, the Glasstron offers private, mobile access to the digital pornoverse: prerecorded, online, and off the Web. The weirdest thing about the Glasstron, though, is the way it configures a doubled voyeurism: the visor is transparent, so you can catch a glimpse in reverse into the user's tele-world. The visor–a prosthetic eye–is not a window on the soul, but rather a real-time display of image consumption (a paradox Situationist Guy Debord did not live long enough to deny as such). This transparency might be useful for panoptical bosses who want to keep an eye on what their workers are keeping their eyes on, I suppose, but it's an odd feature for a consumer electronics product. Thus the lever at the visor's temple: hit it, a shield drops, and the user is left to enhance his solitude in private. Perhaps SONY's engineers were thinking of the bachelors, after all. High-tech, high touch, indeed.

TEOTWAWKI
$
$

TEOTWAWKI

I'M A FAITH VAMPIRE, ONE OF THOSE NON-BELIEVERS OBSESSED WITH BELIEF, AND SO, STARTING IN THE EARLY 1980'S, I BEGAN TO PREPARE NOT FOR THE END TIMES BUT FOR WHAT I'D HOPED WOULD BE AN ABSOLUTE EFFLORESCENCE OF ARCANE PROPHESIES, CHILIASTIC GNOSTICISM, AND GENERAL MILLENNIAL CULTISM.

TEOTWAWKI

With the reaction to the year 2000, my faith in the faith of others evaporated. The world let me down, and hard. I slavered after blood-drenched psychosomatic stigmatae, and was instead offered visions of the Savior on burrito wrappers. None of the usual suspects – atomic war, the Trilateral Commission, the Antichrist – maintained their hold on our entertainment-coddled brains. The bumper sticker of that moment read, "After the Rapture, can I have your car?" Indeed, as we approached the millennium, we feared the acronyms Y2K and TEOTWAWKI, not having any sense of what was really in store for us on 9/11/2K+1.

FIRST, it's worth exhuming and explaining the embarrassing and mostly forgotten Y2K bug hysteria. In the early decades of computing architecture, saving even the smallest amounts of memory was an absolute priority.

Thus, when it came to recording dates, computer languages like Cobol and Fortran halved the amount of space needed to record a year by loping off the indication for the century, so that only two digits, rather than four, were needed: "65" not "1965." Saving two bits may seem insignificant now, but it was a reasonable decision at the time. The problem is that these peculiarities moved with the systems architectures, even when they were upgraded; in the language of the field, they were "grand-fathered in," and so, as the millennial changeover approached, we were faced with an unknown number of machines, all networked together in a vast web, that people feared might shut down or malfunction because they couldn't accept something dated "00" coming after (rather than before) something dated "99." In the techno-apocalyptic imagination, after December 31st, 1999, with the start of the year 2000 (Y2K), computers wouldn't be able to cope, petroleum factories would shut down, food distribution networks would grind to a halt, and financial institutions would face collapse.

FOR THOSE OF US IN THE RICH WEST, THIS WAS A VISION OF:

THE
END
OF
THE
WORLD
AS
WE
KNOW
IT

A WORLD IN WHICH NOT BEING ABLE TO GET CASH NOW FROM THE AUTOMATIC TELLER WOULD SHRED THE VERY FABRIC OF OUT LIVES.

TEOTWAWKI

On one level, the obsession with Y2K was simply an inversion of the era's giddy techno-futurism (which was the strongest wave of optimism since the end of the Space Age). At the height of the boom years, I actually heard pundits and techies say – in public, for God's sake – that the development of the Internet was as important as the discovery of fire to the destiny of humankind. That kind of charmingly loopy optimism by necessity generates its antithesis, a pessimism about all things digital.

Late into the night on talk radio from coast to coast, the voices chimed in hysterical harmony: "Y2K, Millions will die!" Y2K lacked the cachet of a great horned beast breaking seven seals, or even the minute hand of the Doomsday clock on the back cover of the *Bulletin of Atomic Scientists* nudging towards nuclear midnight. How is it that just a few years ago we were reduced to worrying about computer syntax in a world that still had tsunamis, earthquakes, hydrogen bombs and neighbors who hate you with a passion held in check only by the rule of law or the threat of force?

Y2K made manifest a fear that we have created a system not simply beyond our control, but beyond our understanding. The old line has it that all advanced technologies begin as magic, and Y2K implied that they ended that way, as well. Those who professed the Y2K faith took mordant delight when they invoked the end times. Yes, millions would die, but after TEOTWAWKI, the survivors would: a) take a turn to the authoritarian right and embrace patriarchal values (these are the hard-core survivalists, preparing for urban race war); or b) creak towards the ecotopian and live in harmony with Mother Earth (neo-Luddites anticipating technological hubris laid low). The sacrifices of the dead would empower the survivors to purge the system of impious complexities.

Y2K WAS THE last flowering of THE WEST'S 20th century obsession WITH PURITY.

FROM MONDRIAN TO REINHARDT, MODERNISM RESOLUTELY STRIPPED AWAY ALL THAT IT DEEMED SUPERFLUOUS UNTIL THE FORM AND THE CONTENT REACHED DEGREE ZERO.

A WHITE CUBE FILLED WITH WHITE PAINTINGS OFFERS A DIFFERENT AESTHETIC THAN A DIRT-FARMING SURVIVALIST CAMP, BUT BOTH CONSTRUCT ENVIRONMENTS CONSECRATED TO ESSENCE.

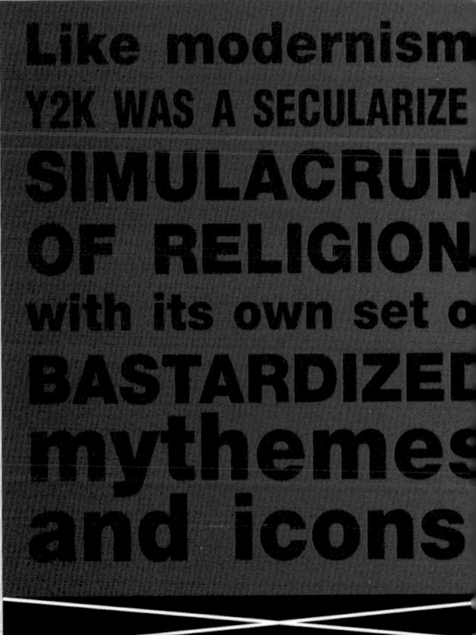

Like modernism
Y2K WAS A SECULARIZE
SIMULACRUM
OF RELIGION
with its own set o
BASTARDIZED
mythemes
and icons

YET WHILE EVEN THE MOST FERVENT CONCEDED THA ITS IMAGINARY WAS IMPOVERISHED (AT LEAST I COMPARISON TO *THE BOOK OF REVELATIONS*), YO DIDN'T HAVE TO BE A BELIEVER TO FEEL SCARE

Atheists are free from the fear of Gog and Magog; Y2K, though, threatened even the unwired technophobe. Those who wanted no part of chat rooms, desktop publishing and server architecture found that unless they were Amish, their lives had been digitized with neither their consent nor their comprehension.

In so many places, information technology (IT) seeps into the fabric of everyday life the way that religion used to. While one could blaspheme in the Middle Ages, it was well-nigh impossible to actually disbelieve in God.

TEOTWAWK

THE FIXATION ON Y2K AT THE EXPE
OF OTHER DOOMSDAY SCENAR
ESPECIALLY THOSE THAT MANIFES
THEMSELVES ON SEPTEMBER 1
2001, WAS A SIGN THAT OUR CULT
IS WAKING UP TO ITS I.T.-DOMINA
PRESENT. THIS GROWING REALIZAT
DIDN'T MAKE ME FEEL ANY BET
ABOUT THE HOME-GROWN BLOODL
OF THE TRUE Y2K FANATICS. Y
ZERO IS SIMPLY TOO STRONG A C
CEPT FOR SOME IMAGINATIONS.
JEWS, CAMBODIANS, AND TUTSIS W
DID EXPERIENCE THE END OF
WORLD AS THEY KNEW IT DURING T
LONG 20TH CENTURY REMIND THE R
OF US THAT ACRONYMS ARE NOT
GREATEST TERROR WE FACE IN T
VEIL OF TEARS.

MASTER LIST

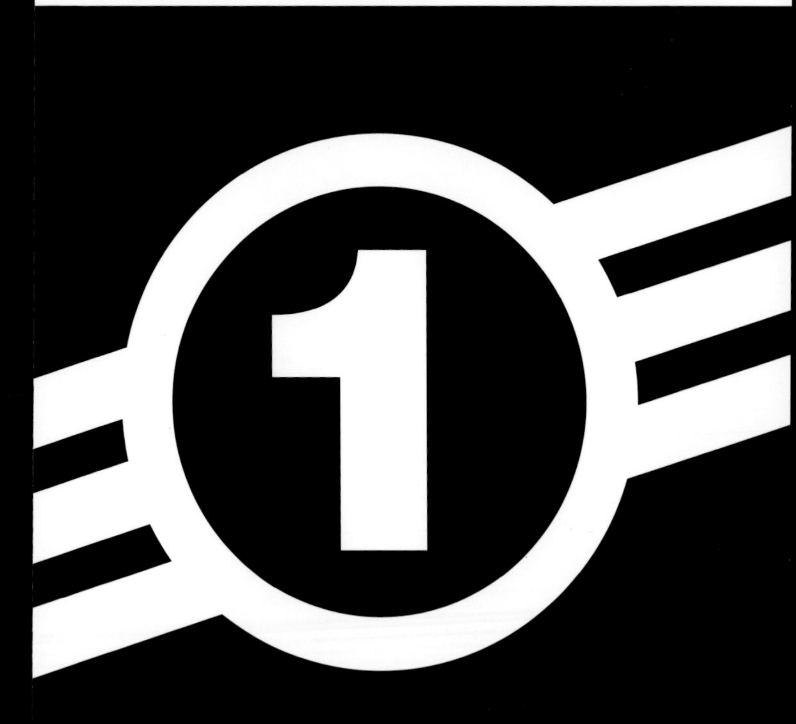

I know a classicist who made a career out of translating the Alexandrine Greek equivalent of dry cleaning lists. These fragments join others in yet another list, his curriculum vitae, that Latin term for the structuring in chronological order of presentations, publications, and positions. I mention this stultifying pedant and his even more stupefying rise to department chairperson not to slag the academic game, but rather to give a sense both of the insignificance and the ubiquity of the list.

MASTER LIST

The list has lately been eclipsed by its more glamorous cousin, the database. There's much talk these days about database art, database narrative, in fact, of an emerging database culture, but outside of set theorists, programmers and information architects, what actually constitutes a database remains fuzzy, even mystical. Relational databases aren't really that complicated. They are simply grids of information, running in rows and columns, and their most powerful attribute is that they can be continuously rearranged to create new relationships from the same set of data. This very fluidity, though, makes databases somehow threatening. Since human beings are reputed to be unable to hold more than seven items in their minds at one time (which was the original rationale for the seven digit telephone number), how are we to negotiate information organized in two dimensions, much less three? The short answer is that we can't, and so we usually request that the database yield something linear rather than spatial: that is, we ask for a list.

MASTER LIST

LISTS SATISFY OUR HUNGER FOR THE UNAMBIGUOUS RANKING OF INFORMATION

The NCAA collegiate basketball tournament winds down from the original field of SIXTY-FOUR teams to the round of the SWEET SIXTEEN then to the much hyped Final Four.

RADIO STATIONS SPIN THE FAVORITE 5 AT 5:00 FROM A PLAYLIST DUTIFULLY BUILT FROM THE TOP 40 AND BILLBOARD'S HOT 100.

The chroniclers of capital compile the richest people as the Forbes 400 and the biggest companies into the Fortune 500.

All of these lists are predicated on the existence of a universally admired Number One, so these are the lists of winners; but anyone who was ever picked last for dodge ball has experienced list building at its rawest and most humiliating.

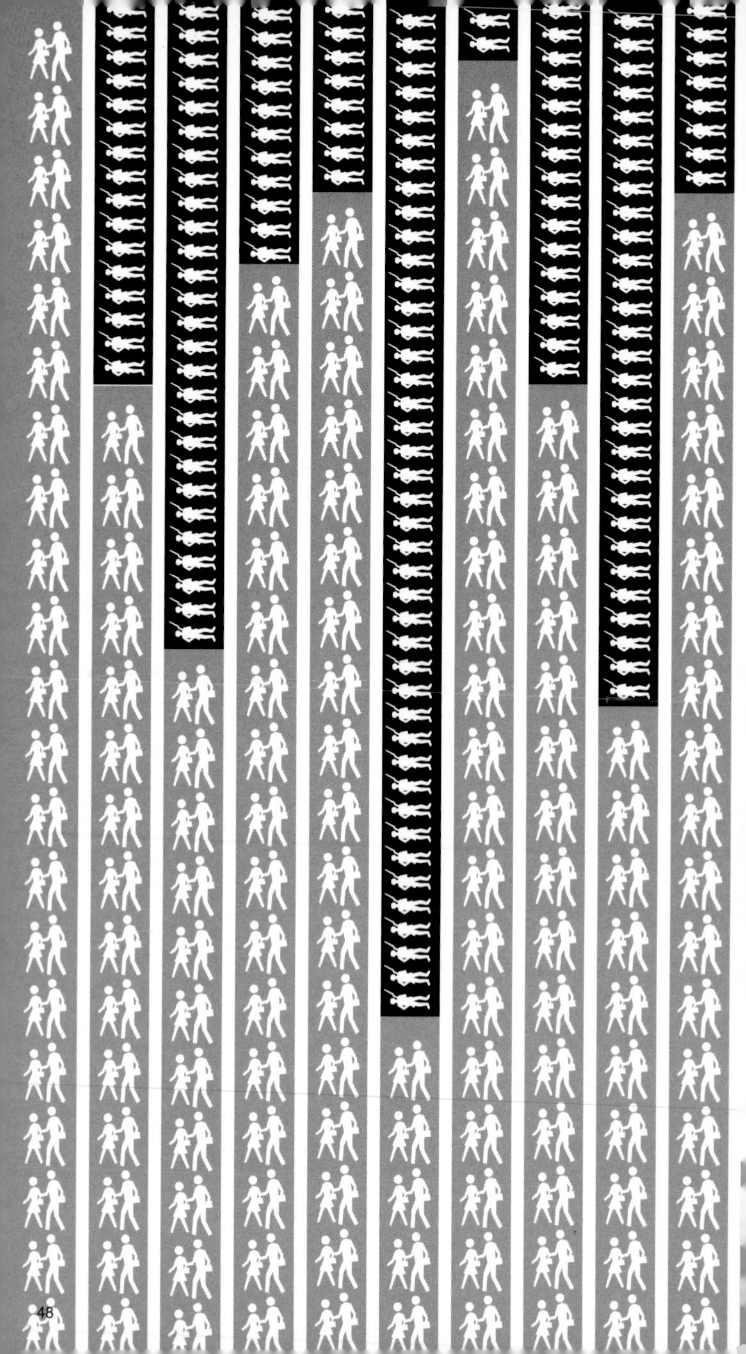

MASTER LIST

LISTS TAKE MANY FORMS

A THERE IS THE DYNAMIC LIST

like Amazon.com's ever-shifting rankings on book sales. Some writers check their place in the pecking order daily, even hourly, in thrall to yet another addiction made possible by online environments.

B THERE ARE THE FINANCIAL LISTS

made possible by these same technologies which transformed markets as electronic investing created a nation of amateur day traders, relentlessly tallying lists of shares to buy, to hold, to sell. The self-absorption of these list makers is breathtaking. At one point before the bubble burst, the most popular shares on E*TRADE, the most popular online investing site, were shares in E*TRADE itself. Thus the tautology of the feedback loop fed the irrational exuberance of the boom years.

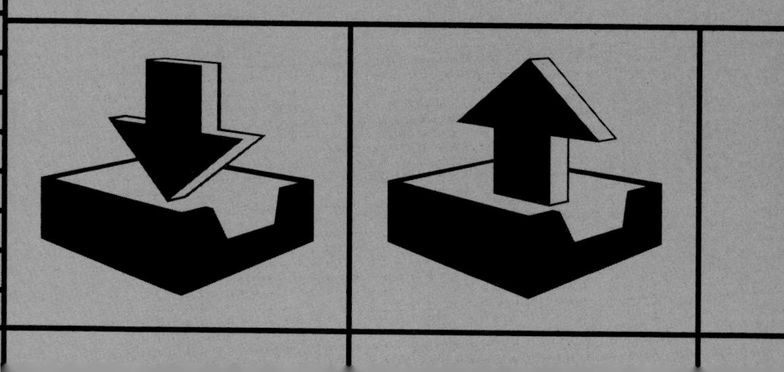

MASTER LIST
THEN THERE IS
THE
TO-DO LIST
THE GREAT
MONUMENT
TO EFFICIENCY
beloved of management
consultants who lead
seminars on the use of
Filofaxes,
Franklin Planners and
Palm Pilots.

MASTER LIST

To-do lists are refined with a reverence reserved for only the most powerful secular rituals. They offer syntactic structure to the martial metaphors of business. The first line of attack for hard-boiled middle managers is to cut the fat, get down to brass tacks, and move on action items. To-do lists are dedicated to the idea that the straightest path is always best. This cult of efficiency moved from business to show business as MBAs in suits replaced the old studio bosses. When he ran Columbia Pictures, Harry Cohn said that he knew a film was good or bad by whether or not his ass burned while he watched it. The pioneering moguls made their share of bombs, but they did believe in telling stories. In his crude way, Cohn attempted a gestalt criticism, an approach now entirely foreign to the anonymous Prada-clad minions who produce today's Hollywood block-busters. These are movies which eviscerate narrative, stripping off the connective tissue of the story until only quick cuts between set pieces remain. The shooting scripts for our biggest budget films do not weave tales, they arrange fights, stunts, explosions, effects, and quips into ever more efficient lists.

THIS IS NOT TO SAY THAT LIST CULTURE DOESN'T HAVE ITS PARTISANS.

MASTER LIST

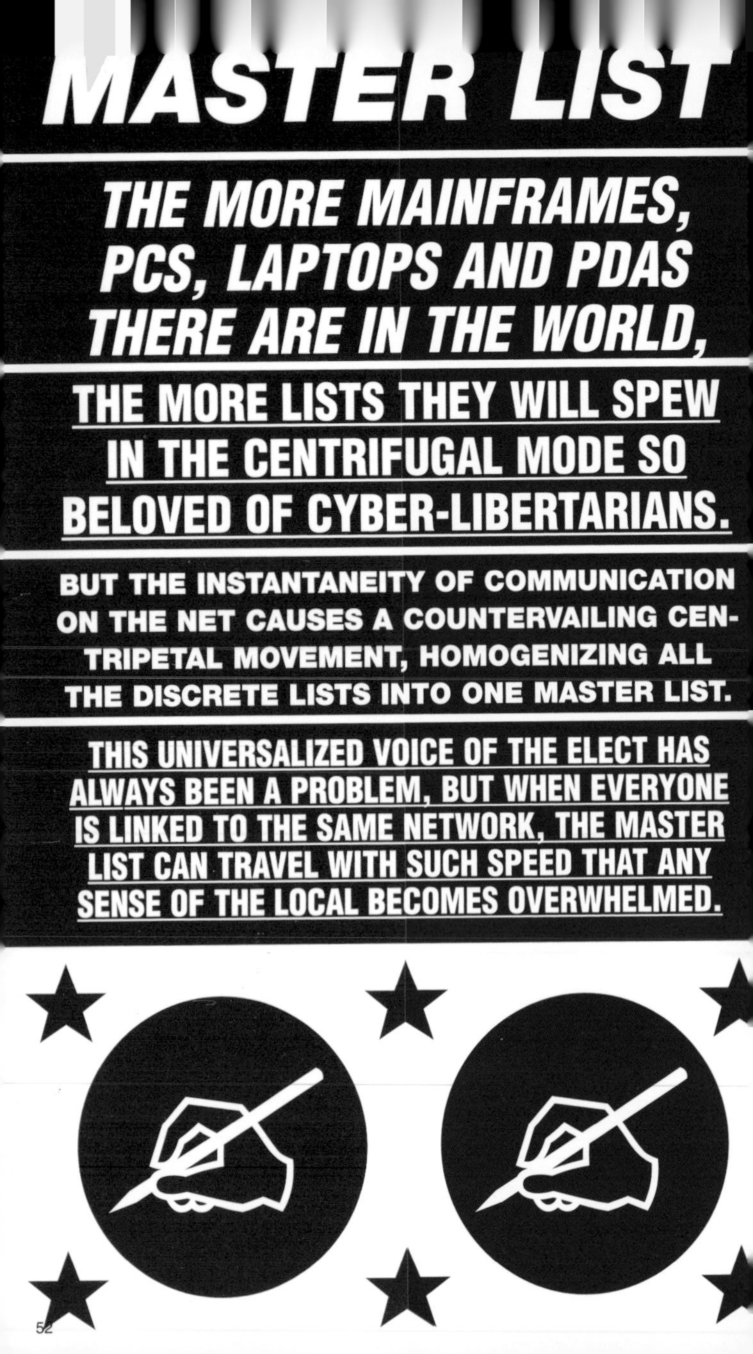

THE MORE MAINFRAMES, PCS, LAPTOPS AND PDAS THERE ARE IN THE WORLD,
THE MORE LISTS THEY WILL SPEW IN THE CENTRIFUGAL MODE SO BELOVED OF CYBER-LIBERTARIANS.
BUT THE INSTANTANEITY OF COMMUNICATION ON THE NET CAUSES A COUNTERVAILING CEN-TRIPETAL MOVEMENT, HOMOGENIZING ALL THE DISCRETE LISTS INTO ONE MASTER LIST.
THIS UNIVERSALIZED VOICE OF THE ELECT HAS ALWAYS BEEN A PROBLEM, BUT WHEN EVERYONE IS LINKED TO THE SAME NETWORK, THE MASTER LIST CAN TRAVEL WITH SUCH SPEED THAT ANY SENSE OF THE LOCAL BECOMES OVERWHELMED.

MASTER LIST

01 02 03 04 05 06 07 08 09 10 11 12 13 14 15 16 17 18 19 20 21 22 23

The rise of international biennales from Seoul to Istanbul to Johannesburg, and the concurrent explosion of jet-setting independent curators, seemed at first to indicate that true globalism and multicultural diversity was at hand for the art world. But look at the catalogues for these mega-shows and you see that a master list is already coalescing. Independent curators trumpet their structural similarities to the peripatetic transnational business class: they eschew their desk-bound institutional colleagues, boasting that they curate from plane seats, and keep in contact via cell phones, hotel fax machines, alphanumeric skypagers and, of course, email. From Rotterdam, R. checks her message service in Paris. H.U. called from Barcelona and leaves his contact information in Sydney, asking her for input for a show he's "making" in São Paulo. R. faxes Australia with her most recent list of hot artists. On the plane to Tokyo, H.U. adds a few names, deletes some others, and emails it to O. in Cairo, who only finds the time to read it when he gets to New York. After tightening up the names still further, O. holds a press conference in Berlin and uses this consensus master list as the curatorial core of his international show, which like all the others he describes as dedicated to celebrating the diversity of the world's art making practice.

MASTER LIST

The new master lists are indeed more inclusive that the master narratives they replaced, but they are already showing the same tendency toward homogeneity. Those who depend on lists to curate should keep in mind Maya Lin's *Vietnam Veterans Memorial*, a work of art and architecture predicated on the balance between the universal and the specific. Visitors without a personal connection to the dead are stunned by the very immensity of the list, by its sheer physicality. The shrunken "V" of the wall allows the eye to wander, scanning first names, scrutinizing last names, and sweeping through the 140 panels that list United States casualties dating from 1959 to 1975. For those whose loved ones' names are inscribed in the black granite, the list shrinks dramatically. It becomes a frame, a stage set, an all-enveloping context for confronting personal loss. The gravity of this list has something to do with its fixity, which serves as a rebuke to the endlessly mutable, yet never dissimilar enough master lists rocketing through the networks.

GAME BOY

On a techno-anthropological outing, a friend once observed an older gentleman playing *Gals Panic II* for three hours straight. *Gals Panic II* is the most deranged video game ever made, a lavishly orchestrated, one-man virtual circle jerk: phallic monsters shoot lethal projectiles at players who are trying to uncover silhouetted figures which when hit, turn into photographs of semi-nude women which themselves transform into the original phallic monsters if the cheesecake doesn't materialize fast enough. The kids at the arcade wandered over anytime they sensed grandpa was about to score a soft-core payoff, blissfully entertained even after their own money had long since run out. The great filmmaker Chris Marker once wrote that video games offer deeper insight into the unconscious than the collected works of Jacques Lacan. I'll buy it. *Gals Panic II* spews Freud's Wolfman as 64-bit twitch candy.

Sure, there have been a few games that have eschewed the formula (those ancient warhorses *Tetris* and *Myst* are usually trotted out at this point, along with the *Sims* franchise), but for the most part, it's blood, mischief, and role playing that gamers revel in. They live in an alternate universe, a solipsistic one scripted by designers whose frame of reference extends no further back than *Pong, Pac-Man* and *Dungeons & Dragons*. The visual and narrative tropes most of us bring with us as cultural baggage are for these teenaged diehards all but forgotten ancestral memories, thrown off in transit, on purpose, too cumbersome to be of any use. Looking over the best-known shooters like *Quake* and *Doom*, and the lesser-known contemporary best-sellers, the titles alone tell you more than you want to know: *Grand Theft Auto, Def Jam Vendetta 2, Warcraft III*. Two years ago, the games numbered "3" were first released; two years from now they'll be selling iteration "4" or "5."

YET ALL OF A SUDDEN, IT SEEMS, THERE ARE PEOPLE OUTSIDE GAMING'S STINKY REC ROOM FOR WHOM ALL THIS MATTERS – A LOT.

Consider the frequent references to gaming in art video, as well as in painting, sculpture, and installation. In part, it's because video games have come to rival the cinema as a contributor to the cultural imagery (economically, they're in a dead heat, game sales and rental income having pulled even with box office receipts). It's also about cultural capital: Lara Croft, the virtual star of the *Tomb Raider* games, burns more brightly than almost any contemporary movie character, so brightly, in fact, that she attracted Gen-X scribe Douglas Coupland to author *Lara's Book*, complete with strategy tips; and the artist Miltos Manetas to create *Flames*, a video of Lara trying to cross a corridor, being struck by arrows, dying, trying again, failing, dying, and so on, in an endless, merry absurd loop. That Hollywood stumbled in adapting her to the big screen not once but twice is only proof of the celluloid dream factory's difficulties in profiting from the emerging digital dreamworks.

GAME BOY

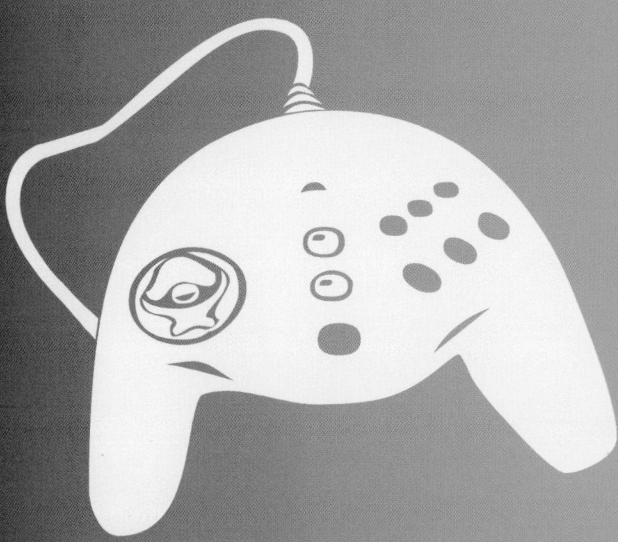

Artists have long been open to games, play, and even sport: think of Marcel Duchamp's obsession with chess; the Surrealists' Exquisite Corpse; the extruded board games that were the Situationists' psycho-geographic mappings of Paris; the algorithmic play of "Oupeinpo" (the Working Group for Potential Painting that sprung up in Oulipo's literary wake); Matthew Barney's appropriation of the double zero from the football jersey of Oakland Raider Jim Otto; and the unforgettable *Vin-da-loo*, the U.K. squad's World Cup anthem co-penned by Damien Hirst.

Today, when an artist like Chris Finely creates suites of paintings with titles referencing

LEVEL THREE and WARP ZONE

you could say that he's taking the classic–and now classically suspect–high road, trying to revitalize or radicalize painting or sculpture with the importation of pop cult tropes.

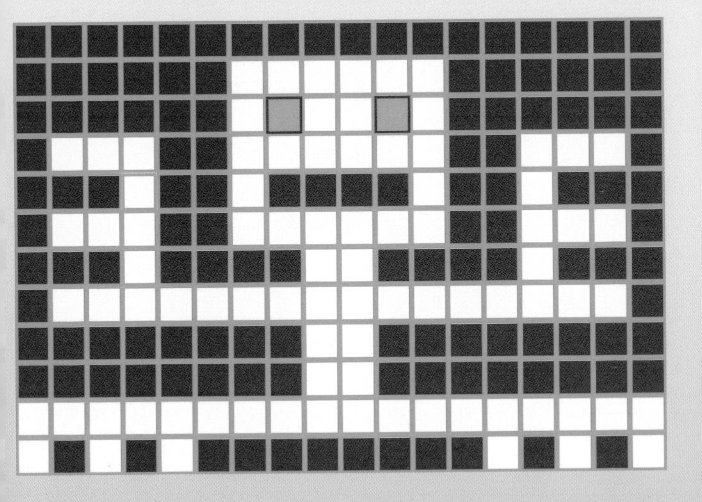

This has been, and still can be interesting, but it can also facilitate a kind of bastardized, twice-emptied-out formalism, with avatars, icons, and the like functioning as yet another brand of geo-metric and/or curvi-linear flourishes.

In the end, though, I think the more convincing, if still subterranean explanation for the increasing fix on the video game mulitverse comes out of the simultaneously exhilarating and creeping feeling that someone, somehow–and sometime soon–is going to completely change the rules.

Could those watching *Mr. Ed*, *Combat* and *Bewitched* in the '60s have guessed that the same technology would be employed to generate Mariko Mori's narcissistic exotica, Tracy Emin's confessionals, or Diana Thater's multicolored theater of high theory? The technology didn't change so much as who controls it. This is not to say I want to see Mori, Emin, Thater, or Bruce Nauman, for that matter, make video games (actually, I think Mori could make amazing ones), but that there is a Game Boy/Girl out there who will grow up and rock our world, for real.

The codes are there for the tweaking. The last time w saw the explosion of a set of narrative, character and stylistic codes was in the 1980s, in comic books. Comics share almost the same demographics as gam ing, at least in North America, and its then-dominant mode of spandex-clad superheroes battling the insatiable forces of doom was mired in a rut that had bee worsening for decades. But with Frank Miller's semin revision of the Batman in *Dark Knight*, Neil Gaiman's reinvigoration of the horror comic in *Sandman*, and Allan Moore's destruction of the Nietzschean superman in *Watchmen*, the '80s supported a mainstream parallel universe where comics resonated with metacritical commentaries on their very subjects and forms. During that period of remarkable creative activ ity, readers had this sense that collective icons and memories were being fractured and recombined to create something novel which nonetheless drew from the same wellspring of generic pleasures that had fed them from pimpledom up. These artists refracted the language and the culture of comics through a new se of sensibilities, creating something that could draw in (at least at that particular moment) a large audience that loved the medium as much as they did, and wanted to see it taken to a new place.

I understood what was going on in comic books two decades ago because, as I'll discuss in a later essay, I grew up pulp, immersed in the codes that were being so tenderly, yet mercilessly manipulated. I doubt I'll be able to identify with the kind of work that artists seem to be sensing is just around the corner in gaming because it will inflect a language and a culture that has been peripheral to my adult life. But then again, the kind of work I'm envisioning won't be made for me. Maybe when the Game Boy grows up, he'll realize that interesting work isn't necessarily interesting to everyone.

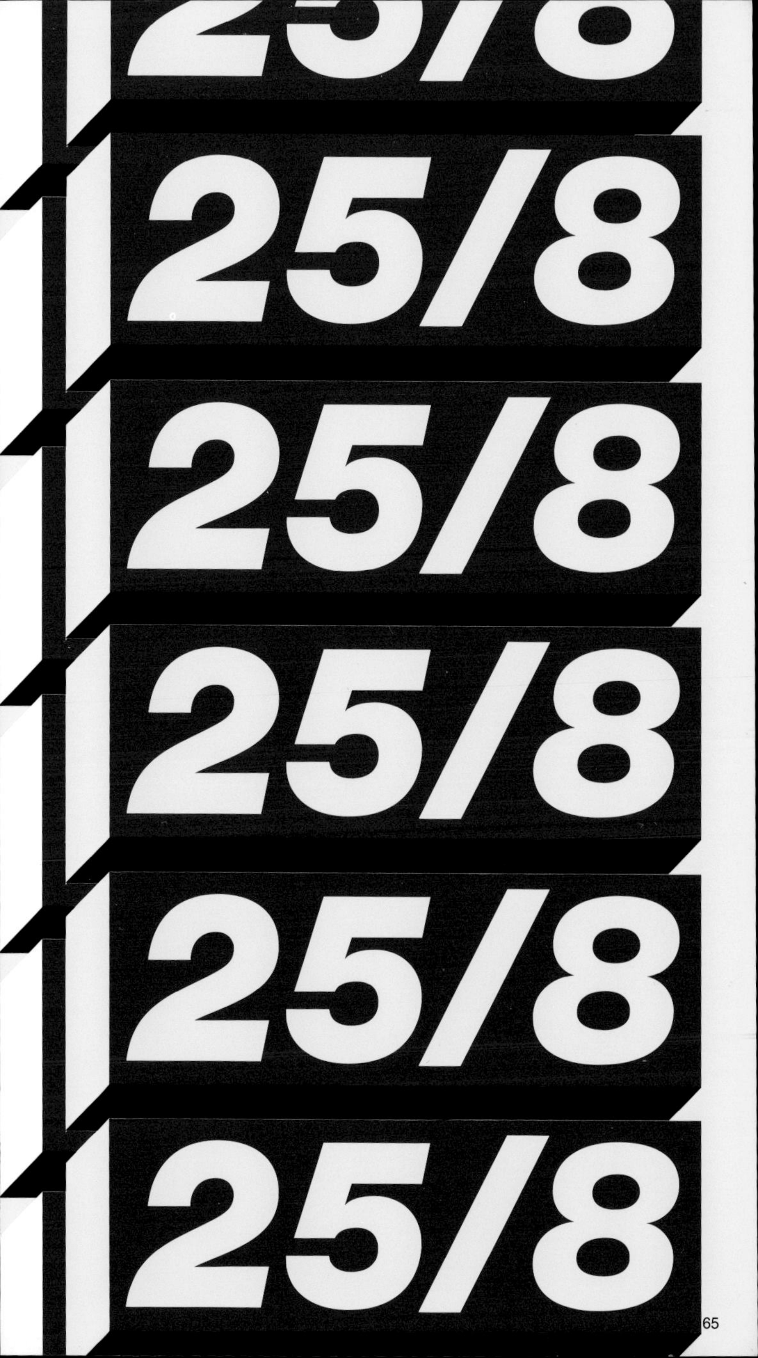
25/8
25/8
25/8
25/8
25/8
25/8

25/8-25/8-25/8-25/

CELEBRITIES WITH ONE NAME ARE EASIER TO REMEMBER, AND HENCE HARDER TO FORGET, I.E. CHER, HALSTON, EMINEM.

SO WHEN SLASH, VELVET REVOLVER'S GUITARIST BEST REMEMBERED FROM HIS STINT IN GUNS N' ROSES, WAS INTERVIEWED ON THE RADIO, I CEDED HIM BITSPACE SIMPLY BECAUSE HE DOESN'T HAVE A LAST NAME. SLASH TALKED ABOUT HOW HARD HE'D WORKED SCORING A CABLE TV MOVIE ABOUT PORNOGRAPHERS, AND CLAIMED TO HAVE HIT THE EDGE OF EXHAUSTION "WORKING 25/8."

My faith in the UNI-*NOMENKLATURA* paid off handsomely. Here was the tired cliché of the hard-charging entrepreneur working 24 hours a day, 7 days a week raised to new heights, like the lawyer I know who bills 27 hours in a single calendar day because he includes the time change on his flight from New York to L.A.

5/8-25/8-25/8-25/8

SO MANY PEOPLE LAY CLAIM TO THE MANTLE OF 24/7 THESE DAYS THAT IT'S NO MORE MEANINGFUL THAN THE ENDLESS STREAM OF MIDDLE MANAGERS EXHORTING THEIR UNDERLINGS TO "THINK OUT OF THE BOX," THEREBY CREATING A WHOLE NEW BOX.

25/8
IS STILL
FRESH
ENOUGH
TO BE A LEAP
OF SELF-
DESIGNATED
BUSYNESS

5/8-25/8-25/8-25

25/8 is the inversion of the Aristotelian maxim that the unexamined life is not worth living. Someone living 25/8 perforce has no time to examine anything at all. 25/8 is the human trying to push past the limits of the flesh into the real of pure performance, like an engine or a chip. Financial networks meld Tokyo, London and New York into a 24/7 global market, explaining if not excusing the desire to go an hour longer and a day better. It's easy enough to pass judgment on the 25/8 lifestyle–the TV executive who rises every morning at 4:30 to get a jump on his competitors simply to pump ever worse crap out on the airwaves–but let's remember that 25/8 is a symptom, not a cause. Historians of the book periodize the first 50 years after Gutenberg as the age of the *incunabula*, Latin for "swaddling clothes." With hundreds of magazines, thousands of channels, millions of books, billions of Web pages, and what only seems like trillions of blogs, it's not inaccurate to refer to our present millennial moment as one of digital incunabula, with the cult and culture of information undergoing an epistemic big boom

25/8 IS YET ONE MORE PROOF OF THE COMPLETE VICTORY OF THE DROMOCRACY, THE MONARCHY OF SPEED. SPEED IS NOT ONLY TRIUMPHANT, IT IS VIRTUALLY UNCHALLENGED. SURE, MEDIA CONGLOMERATES LAUNCH THE OCCASIONAL MAGAZINE WITH TITLES LIKE *REAL SIMPLE* (OR SOME SUCH OTHER BLAND EXHORTATION TO SPEND YOUR WAY TO TRANQUILITY), AND *FOR COMMON THINGS* (A BOOK ABOUT WALKING THOREAUVIAN BEFORE GOING OFF TO LAW SCHOOL BY THE PROUDLY NAÏVE YOUNG HARVARD GRAD JEDEDIAH PURDY) WAS A BESTSELLER, BUT IT DOESN'T LOOK LIKE THE TWENTY-FIRST CENTURY WILL DECELERATE ONE BIT. THAT SAID, THE ISSUE BECOMES HOW INDIVIDUALS AND INSTITUTIONS WILL NEGOTIATE THEIR FATES AND AMBITIONS UNDER THIS REGIME.

25/8-25/8-25/8-25/

IN OTHER WORDS,

BEYOND SIMPLY PROCLAIMING THE IMPOSSIBLE GOAL OF LIVING

5/8-25/8-25/8-25/8

NALIZING

VELOCITY?

25/8-25/8-25/8-25/

RATHER THAN FRETTING OVER OVER-COMMITMENT IN THE ERA OF 25/8 SPEED UP, A SELECT GROUP HAS DECIDED THAT GIVEN OUR SUPERSATURATED, EVER-ACCELERATING SOCIETY, THE SOLUTION IS TO EMBRACE THE MILLENNIAL PROMISE OF BEING EVERYTHING, EVERYWHERE, ALL THE TIME. JET PLANES OBLITERATE PHYSICAL DISTANCES, NETWORKS CREATE VIRTUAL ENVIRONMENTS, AND AMBITIONS DILATE ACCORDINGLY.

5/8-25/8-25/8-25/8

So it is that we now live in an era that claims to be a new renaissance, in which

actors can be **singers, singers** strive to be **artists, painters** become **film directors, digital artists** say that they are **scientists, scientists** become **entrepreneurs, entrepreneurs** wake up one morning thinking they are **politicians,** and **politicians,** well, they've always been so protean (folksy at home, regal in the statehouse) that they are **the poster children for the millennially ambitious.**

Too many blindly recapitulate the imaginary of Warhol's Factory, where desire mattered more than talent.

WHO AM I TO DENY ANYONE THEIR DREAM OF BEING THE NEW LEONARDO?

(Still, the fact remains that the only Renaissance man I know works at a Renaissance faire.)

WHEN INDIVIDUALS FALL INTO THIS FANTASY, THE DAMAGE IS LOCAL-IZED, THE HUMILIATIONS PERSON-AL, AND FOR THAT FEW WHO SUC-CEED, THE ENORMOUS PAYOFFS MAKE THE MORTIFICATIONS ALONG THE WAY PALATABLE.

5/8-25/8-25/8-25/8

WHEN INSTITUTIONS FOLLOW INDIVIDUALS

into the halogen-lit realms of 25/8, however,

THE EMBARRASSMENTS ARE MORE PUBLIC.

It's been decades now, for example, since museums knew what they were, ontologically speaking. Your King Tut and Manet and Picasso blockbusters have forever skewed the notion of audience. Ever-expanding, never-contracting education departments would rather have no object at all than give up their dithering didacticism, matters made even worse as noisy multimedia experiences supplant explicatory wall texts. The very question of what exactly museums should show is up for grabs: the Guggenheim – always at the forefront of museology, if not tastemaking – exhibits everything from BMW motorcycles to Norman Rockwell. There's also the inevitable boredom factor to account for. Who wants to mount "Masters of Impressionism XVII"?

25/8-25/8-25/8-25/

Some time back, over the course of a single week, I saw three major survey shows on two coasts: the inaugural net.art selection at the Whitney Biennial, the Cooper Hewitt's Design Triennial, and a massive survey at LA MOCA, "At the End of the Century: One Hundred Years of Architecture." Whitney Biennials are like Roman candles, bright and exciting when they first start, with a long slow fizzle at the end. I finally saw the show during its final week, long after the New York art world had definitively spoken on which bad paintings from Texas got the museum's primo wall space, etc., etc. I was particularly interested in determining whether the museum had spent as much time thinking about how to show net.art as they did promoting the fact that it was included in the Biennial in the first place. The answer was disappointing, as the net.arts were relegated to the ghetto of a projection room, with a single computer/mouse station at the rear, so that one user–inevitably making this his or her day's activity–could drive the experience for everyone else. This display strategy had been discarded by serious curators of media arts five or six years before–in Internet time an unforgivably distant past. In this case, the Whitney was a museum wanting to be a monitor, and failing badly.

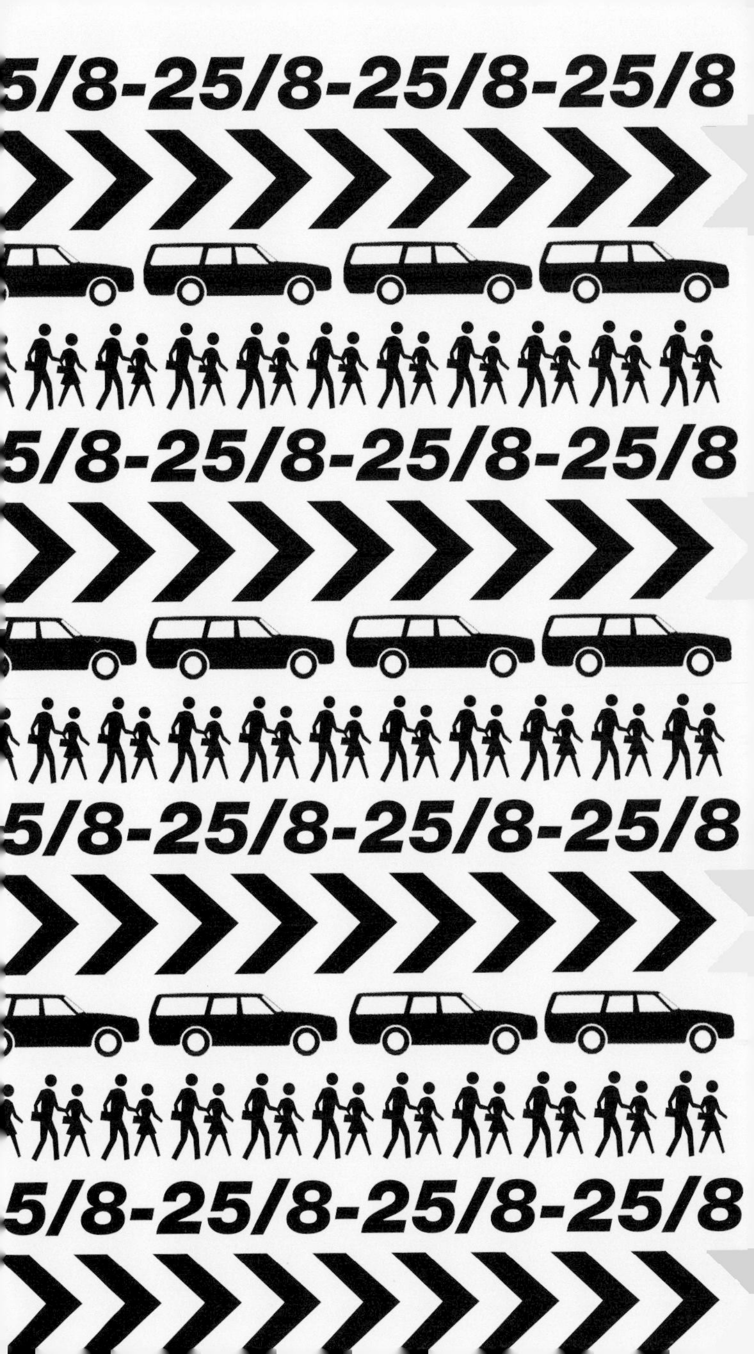
5/8-25/8-25/8-25/8
5/8-25/8-25/8-25/8
5/8-25/8-25/8-25/8
5/8-25/8-25/8-25/8

25/8-25/8-25/8-25/

The Design Triennial at the Cooper-Hewitt was vibrant and novel in comparison, but encasing the blobjects of commercial culture in vitrines, the show put out of reach that which was specifically about being for sale. The Triennial was a store masquerading as a show. "At the End of the Century" (AEC) was the best-curated and most compelling of the three shows (it had 100 years to cull from, of course, rather than two or three), but its very vastness made me wonder where its ambitions ended. AEC needed so much wall text to explain architectural history that it approached the limits of a show catalogue. The computer-generated 3-D walkthroughs of unbuilt structures, like Tatlin's *Monument to the Third International*, lent parts of AEC the air of a theme park for architectural play. But in the end, the miniaturized maquettes throughout the echoing spaces of Frank Gehry's Geffen Contemporary left me feeling that AEC aspired to the very condition of architecture itself: an exhibition that wanted to be a building. Somehow, immersing myself in these three shows reminded me of Damien Hirst's sumptuously printed, Jonathan Barnbrook designed paean to internalized velocity, *I Want to Spend the Rest of My Life Everywhere, With Everyone, One to One, Always, Forever, Now.* How very 25/8.

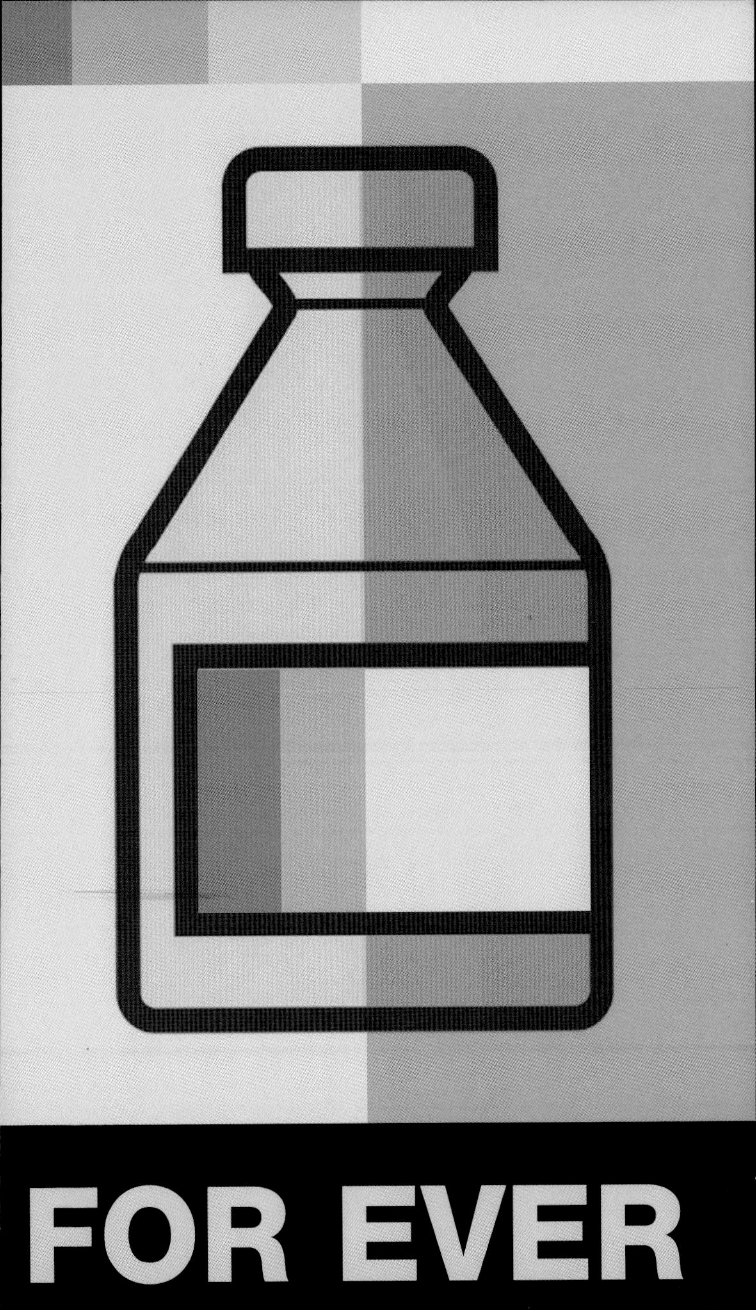
FOR EVER

FOR EVER

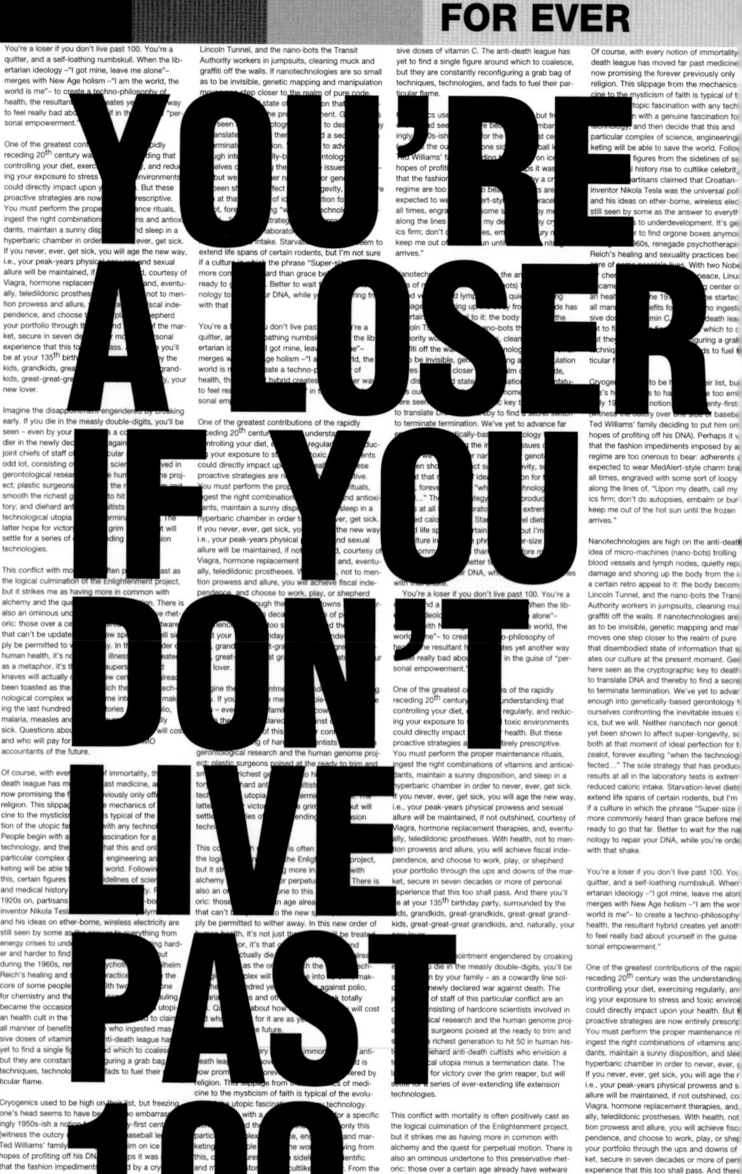
YOU'RE
A LOSER
IF YOU
DON'T
LIVE
PAST
100.

Nanotechnologies are high on the anti-death list. The idea of micro-machines (nano-bots) trolling through blood vessels and lymph nodes, quietly repairing damage and shoring up the body from the inside has a certain retro appeal to it: the body becomes the

Reich's healing and sexuality practices became the core of some people's lives. With two Nobels, one for chemistry and the other for peace, Linus Pauling became the occasionally unwilling center of a utopian health cult in the 1970s when he started to claim all manner of benefits for those who ingested mas-

ing the last hundred years' victories against polio, malaria, measles and other scourges look totally sick. Questions about how much immortality will cost and who will pay for it are as yet left to HMO accountants of the future.

YOU'RE A QUITTER, AND A SELF-LOATHING NUMBSKULL. WHEN THE LIBERTARIAN IDEOLOGY *– I GOT MINE, LEAVE ME ALONE –* MERGES WITH NEW AGE HOLISM *– I AM THE WORLD, THE WORLD IS ME –* TO CREATE A TECHNO-PHILOSOPHY OF HEALTH, THE RESULTANT HYBRID CREATES YET ANOTHER WAY TO FEEL REALLY BAD ABOUT YOURSELF IN THE GUISE OF *"PERSONAL EMPOWERMENT."*

One of the greatest contributions of the rapidly receding 20th century was the understanding that

controlling your diet, exercising regularly, and reducing your exposure to stress and toxic environments could directly impact upon your health.

But these proactive strategies are now entirely prescriptive

You must perform the proper maintenance rituals, ingest the right combinations of vitamins and antioxidants, maintain a sunny disposition, and sleep in a hyperbaric chamber in order to never, ever, get sick. If you never, ever, get sick, you will age the new way, i.e., your peak-years physical prowess and sexual allure will be maintained, if not outshined, courtesy of Viagra, hormone replacement therapies, and, eventually, teledildonic prostheses. With health, not to mention prowess and allure, you will achieve fiscal independence, and choose to work, play, or shepherd your portfolio through the ups and downs of the market, secure in seven decades or more of personal experience that this too shall pass.

AND THERE
YOU'LL BE AT YOUR
135th BIRTHDAY PARTY,
SURROUNDED BY
THE KIDS, GRANDKIDS,
GREAT-GRANDKIDS,
GREAT-GREAT GRAND-
KIDS, GREAT-GREAT-
GREAT GRANDKIDS,
AND, NATURALLY,
YOUR NEW LOVER.

IMAGINE THE DISAPPOINTMENT
ENGENDERED BY CROAKING EARLY.
IF YOU DIE IN THE MEASLY DOUBLE-
DIGITS, YOU'LL BE SEEN – EVEN BY
YOUR FAMILY – AS A COWARDLY LINE
SOLDIER IN THE NEWLY DECLARED
WAR AGAINST DEATH. THE JOINT
CHIEFS OF STAFF OF THIS PARTICU-
LAR CONFLICT ARE AN ODD LOT,
CONSISTING OF HARDCORE SCIEN-
TISTS INVOLVED IN GERONTOLOGICAL
RESEARCH AND THE HUMAN GENOME
PROJECT; PLASTIC SURGEONS POISED
AT THE READY TO TRIM AND SMOOTH
THE RICHEST GENERATION TO HIT 50
IN HUMAN HISTORY; AND DIEHARD
ANTI-DEATH CULTISTS WHO ENVISION
A TECHNOLOGICAL UTOPIA MINUS A
TERMINATION DATE.
THE LATTER HOPE FOR VICTORY OVER
THE GRIM REAPER, BUT WILL SETTLE
FOR A SERIES OF EVER-EXTENDING
LIFE EXTENSION TECHNOLOGIES.

This conflict with mortality is often positively cast as the logical culmination of the Enlightenment project, but it strikes me as having more in common with alchemy and the quest for perpetual motion. There is also an ominous undertone to this preservative rhetoric: those over a certain age already have wetware that can't be updated to the new specs, and will simply be permitted to wither away. In this new order of human health, it's not just that illness will be treated as a metaphor, it's that only paupers, fools and knaves will actually die. Our new century has already been toasted as the one in which the medico-technological complex will truly come into its own, making the last hundred years' victories against polio, malaria, measles and other scourges look totally sick. Questions about how much immortality will cost and who will pay for it are as yet left to HMO accountants of the future.

Of course, with every notion of immortality, the anti-death league has moved far past medicine, and is now promising the forever previously only offered by religion.

THIS SLIPPAGE FROM THE MECHANICS OF MEDICINE TO THE MYSTICISM OF FAITH IS TYPICAL OF THE EVOLUTION OF THE UTOPIC FASCINATION WITH ANY TECHNOLOGY.

People begin with a genuine fascination for a specific technology, and then decide that this and only this particular complex of science, engineering and marketing will be able to save the world. Following from this, certain figures from the sidelines of scientific and medical history rise to cult-like celebrity. From the 1920s on, partisans claimed that Croatian-born inventor Nikola Tesla was the universal polymath, and his ideas on ether-borne, wireless electricity are still seen by some as the answer to everything from energy crises to underdevelopment. It's getting harder and harder to find orgone boxes anymore, but during the 1960s, renegade psychotherapist Wilhelm Reich's healing and sexuality practices became the core of some people's lives. With two Nobels, one for chemistry and the other for peace, Linus Pauling became the occasionally unwilling center of a utopian health cult in the 1970s when he started to claim all manner of benefitsfor those who ingested massive doses of vitamin C. The anti-death league has yet to find a single figure around which to coalesce, but they are constantly reconfiguring a grab bag of techniques, technologies, and fads to fuel their particular flame.

Cryogenics used to be high on their list, but freezing one's head seems to have become too embarrassingly 1950s-ish a notion for the twenty-first century (witness the outcry over one side of baseball legend Ted Williams' family deciding to put him on ice in hopes of profiting off his DNA). Perhaps it was simply that the fashion impediments imposed by a cryonics regime are too onerous to bear: adherents are expected to wear MedAlert-style charm bracelets at all times, engraved with some sort of loopy message along the lines of, "Upon my death, call my cryogenics firm; don't do autopsies, embalm or bury me, and keep me out of the hot sun until the frozen nitrogen arrives."

ANOTECHNOLOGIES ARE HIGH ON THE ANTI-DEATH LIST. IE IDEA OF MICRO-MACHINES (NANO-BOTS) TROLLING IROUGH BLOOD VESSELS AND LYMPH NODES, QUIETLY EPAIRING DAMAGE AND SHORING UP THE BODY FROM IE INSIDE HAS A CERTAIN RETRO APPEAL TO IT:

THE BODY BECOMES THE LINCOLN TUNNEL, AND THE NANO-BOTS THE TRANSIT AUTHORITY WORKERS IN JUMPSUITS, CLEANING MUCK AND GRAFFITI OFF THE WALLS.

If nanotechnologies are so small as to be invisible, genetic mapping and manipulation moves one step closer to the realm of pure code, that disembodied state of information that so infatuates our culture at the present moment. Genotech is here seen as the cryptographic key to death—the way to translate DNA and thereby to find a secret switch to terminate termination. We've yet to advance far enough into genetically-based gerontology to find ourselves confronting the inevitable issues of eugenics, but we will. Neither nanotech nor genotech has yet been shown to affect super-longevity, so they are both at that moment of ideal perfection for the zealot forever exulting "when the technology is perfected..." The sole strategy that has produced any results at all in the laboratory tests is extremely reduced caloric intake. Starvation-level diets seem to extend life spans of certain rodents, but I'm not sure if a culture in which the phrase "Super-size it!" is more commonly heard than grace before meals is ready to go that far. Better to wait for the nanotechnology to repair your DNA, while you're ordering fries with that shake.

VISUAL INTELLECTUALS

EVEN WHEN INTELLECTUALS AREN'T TALKING ABOUT WORDS, THEY EXPRESS THEMSELVES THROUGH, BY AND WITH THEM TO SUCH AN EXTENT THAT WHAT THEY GENERATE CAN NEVER TRULY BE SEEN AS A DISCUSSION ABOUT ANYTHING OTHER THAN THEM.

This explains the way that no matter how much art history, design criticism, or new media studies claim to deal with visual and spatial systems, these discursive modes tend to resolve themselves finally around, well, around discourse itself. This is not to say that this text-based intellectual work is in the end consecrated to the craft of writing, as anyone who has valiantly pushed through reams of turgid academic prose can attest (word processing aside, Truman Capote's chestnut, "That's not writing, that's typing," applies). But something new is brewing. I would claim that we are about to witness the wide-scale emergence of visual intellectuals–people simultaneously making, pondering and commenting on visual culture, but in a way that doesn't perforce adhere to the primacy of the word.

VISUAL INTELLECTUALS

These are the people creating the visual culture that surrounds us, a culture that over the course of the past hundred years has essentially supplanted text's preeminence. It would be easy enough to write this development off with the cliché about the triumph of the mute image over the expressive word. But we're long past that narrative now, willing to lift our "downcast eyes" (to cite historian Martin Jay) to look into the light box. Although the utopian promise that will allow people to "write" with audiovisual media often recedes with each new advance (at least on the part of the dominant broad- and even narrowcasters), there is a growing body of work that proves that complex argumentation, sophisticated critique, and even languages of praise are being generated outside of purely text-based discourse.

The World Wide Web is the obvious place to go looking for such multi-mediated ways of thinking. This is, after all, a medium in which the object, that of which it is composed (the source code), and any commentary on that object all exist contemporaneously and conceptually in the same place/non-place of the network.

THE ABILITY TO SCALE WINDOWS WITHIN WINDOWS,

TO CREATE INSTANTANEOUS LINKAGES,

AND TO COMMENT ON THE DEVELOPMENT OF AN ART MOVEMENT USING AN IDENTICAL MODE OF PRODUCTION AND DISTRIBUTION

ALL OF THIS HAS LED TO THE PARTICULAR FLAVOR OF VISUALIZED, HYPER-CODED META-COMMENTARY.

THE FIRST SUCH INSTANTIATIONS WERE ADMITTEDLY SOPHOMORIC–I.E., SITES LIKE SUCK.COM (WEB PAGES THAT SUCK, GET IT?)

– BUT THINGS *IMPROVED AS THE NET.ARTS EVOLVED,* AND IT BECAME OBVIOUS THAT THE ART AND THE DISCOURSE ABOUT THAT ART WERE CONTEXTUALLY AND CONSTITUTIVELY INDISTINGUISHABLE.

THERE WAS ALSO A WILLINGNESS TO EXPLORE META-STRUCTURING OF DATA AS ART, AS WITH *IOD'S* REMARKABLE DECONSTRUCTION DEVICE, *WEBSTALKER*,

Lisa Jevbratt's SOFTWARE ART PROJECT 1:1, A VISUALIZED MAPPING OF THE WEB,

OR TO USE THE STRUCTURES OF THE DIGITAL MEDIA TO ACTIVELY INTERVENE INTO LONGSTANDING DEBATES,

as with pseudo-gaming model of Lev Manovich and Norman Klein's *Freud-Lissitzky Navigator*.

Too often, though, the Web breeds a techno-solipsism, an unwarranted confidence that computer networks are generating something entirely without precedent.

This is nonsense, of course, as avant-garde film and video offer a long history of audio-visual essays and meta-critical production.

To name just a few: in the '60s, there wasTom Tom the Piper's Son, Ken Jacobs's reframing at varying speeds and in different sections of an example of early cinema; in the '70s, Joan Jonas's structuralist video intervention Vertical Roll ; in the '80s, the hyper-theorizations of Gary Hill's proto-interactive single channel piece, Site Recite: A Prologue; and in the '90s, Trouble in the Image, Pat O'Neil's magnum opus of cinematic optical printing.

VISUAL INTELLECTUALS

Perhaps better understood by thinking netizens is the debt to graphic design. Although some ardent youngsters (and not-so-youngsters, unfortunately) protest that something as commercially "tainted" as the professional practice of design has nothing to say to artists like themselves, the impact of contemporary graphics is indisputable. While modernist masters like Paul Rand promoted the ideal of the designer as refining reagent, the substrate through which someone else's message could be filtered, contemporary designers no longer feel obliged to make a show of such modesty. Rand's model was already being dismantled when desktop publishing exploded, radically dropping the price of sophisticated visualization tools (programs like Photoshop, Fontographer and Pagemaker) and fostering an efflorescence of style for style's sake. More self-conscious designers also woke up to the complex challenges to "clarity" accumulating under the rubric of postmodern theory, and began to conceptualize how digital technologies could allow them to develop their own signature styles. The best architectural publications have long been examples of visual intellectuality, and the 1,344-page collaboration between architect Rem Koolhaas and designer Bruce Mau that is *S,M,L,XL* was rightly lauded. In *S,M,L,XL*, the point is neither to illustrate words not to caption pictures, but rather to create a synergistic matrix of images and texts.

No one has ever accused the painter Peter Halley of an inability to read the zeitgeist, so it's instructive to see how he has reacted to the emergence of visual intellectuality. Back in the '80s, Galerie Bruno Bischofberger published *Peter Halley: Collected Essays 1981-87* (designed by Anthony McCall Associates) which, like the Semiotext(e) books so popular at the time, was an elegant, understated, monochromatic text that announced its seriousness and modesty to the point of having a brown paper cover. Contrast that approach to the overwhelming seduction of the recent *Peter Halley: Maintain Speed*. Edited by Halley's studio director Corey Reynolds and designed by COMA (Cornelia Blatter and Marcel Hermans, who also worked on Halley's magazine *Index*), *Maintain Speed* is a Peter Halley production from exploding pink cover to incredibly detailed colophon. Like *S,M,L,XL*, it offers a new (if expensive) model for the visual intellectual.

One of the things that distinguishes this volume from other catalogues is that the reproductions of the paintings, the installation shots, and the incidental photography of the artist and his milieu are all subtended by a delirious grid of parenthetical and relational databases.

MANY OF THE PAINTINGS ARE, OF COURSE, GRIDS, SO THERE IS AN IMMEDIATE RELATIONSHIP BETWEEN THE CONTENT AND THE FORM.

Continuing this is a motif which the editorial and design team referred to informally as the "information bar": a row of ten, postage-stamp-sized boxes, delineated by pink, perforated lines, running along the bottom of each page. These "stamps" are filled only occasionally, sometimes in blocks of two or three, and can be images, diagrams, captions, or quotes. This allows not for a single parallel, but a multiplicity of argumentations and contextualizations of the work under discussion.

This strategy is taken to its utmost when the rows of stamps become pages of them, with a ten-by-ten grid of the pink, perforated lines defining the field for a postage stamp-sized Halley retrospective. The first double-page spread is devoted to the year 1981, and features just five paintings on one page and two on the other. By the time you get four spreads deeper, those original seven paintings have been augmented by 42 more, and their spatial relationships have remained consistent, though they' ve been compacted together toward the Y-axis. Turn the page, however, and the planar development of chronological sequencing is challenged abruptly. A series of blue, curvilinear arrows is overprinted on the exact same grid from the previous spread, but this time creating a flow chart that indicates the conceptual and stylistic linkages both forwards and backwards in time. Along with the subsequent spreads, this offers as beautiful a double mapping of the diachronic and synchronic (to appropriate the theoryspeak of which Halley was so enamored in the '80s) as you are likely to see anywhere.

By discussing *S,M,L,XL* and *Peter Halley: Maintain Speed*, I've obviously stacked the decks, as these projects were masterminded respectively by an architect and an artist. How might historians of music, political scientists, demographers, or feminist legal scholars spin their tales, fabri- [illegible] or illuminate their causes in commensurately dynamic ways?

[illegible] from the mire of formulaic [illegible] with which we are [illegible] point of sheer apathy?

Perhaps first we need to let go of the notion that language is the sober way to truth, and put the visual's intoxicating powers to use doing something other than selling sex, stuff, or (as with so much of today's art and design) simply itself.

URINE
NATION

I'm pretty sure *that all* LANGUAGES and SIGN SYSTEMS will be unified in our lifetimes,

the only thing is, this communicative utopia will be based on Pissing Calvins.

✝✝✝✝✝✝✝✝✝✝✝✝

URINE NATION

Those of you who don't live in Texas, go to NASCAR races, or cruise around in mini-trucks may be unfamiliar with Pissing Calvin, the unauthorized micturating knockoff of Bill Watterson's comic strip, *Calvin and Hobbes* (the Calvin and Hobbes in question being a little kid and a friendly talking tiger rather than a couple of famously dour thinkers).

Watterson never licensed his characters, but this didn't stop, and may indeed have fueled the ever-expanding desire for stickers of Calvin, back turned to the spectator, leaning around to flash his mischievous grin while pissing an arc of urine onto something. This something can be literally anything. Car culture spawned the original pissing match. Ford owners bought Calvin pissing on the Chevrolet badge, Chevy dudes pasted Calvin leaking on Ford's logo. Then the floodgates opened: Calvin pissing on the words, "MY EX-WIFE" or "MY EX-HUSBAND"; Calvin in a Mexican sombrero pissing on "La Migra"; Calvin pissing on an ankh (still haven't figured that one out); even the endlessly recursive bumper sticker of Calvin pissing on a bumper sticker of Calvin pissing on a bumper sticker of Calvin pissing

URINE NATION

THERE ARE

NOW, SQUATTING WHERE THEIR
MALE COUNTERPART ONCE STOOD,

PATRIOTIC
CALVINS

PISSING ON “TERRORISTS,”
AND SIGHTED IN AN MIT CAFÉ –

PISSING ON THE EARTH ITSELF.

URINE NATION

IF YOU HAVE SOMETHING YOU WISH TO COMMUNICATE IN PISSING CALVIN LA
GUAGE, JUST LOG ONTO *CUSTOMVEHICLEGRAPHICS.COM* AND "PERSONALIZE"
THE ARKANSAS-BASED COMPANY'S WORDS) YOUR CALVIN PISSING ON "A PARTIC
LAR PERSON, MOOD, ATTITUDE, SPORTS EVENT, BUSINESS OR SITUATION."

IF YOU LIVED IN YOUR CAR, YOU'D BE HOME BY NOW
IF YOU WANT WORLD PE
FIGHT FOR JUSTI
MY JOB IS SECURE
NO ONE ELSE WANTS I
THIS IS MY OTHER CAR!
AND ON THE EIGHTH DA
GOD WENT FISHI
PAY NO ATTENTION TO THE MAN BEHIND THE CURTAIN

DON'T STEAL
THE GOVERNMENT HATES COMPETITION

GUN CONTROL ISN'T ABOUT GUNS. IT'S ABOUT CONTRO

SAVE THE HUMANS

I'M NOT ALWAYS RIGHT,
BUT I'M NEVER WRONG!

WE ARE BORN NAKED, WET AND HUNGR
THEN THINGS GET WORS

I SOUPORT PUBLIK EDUKASHU

WE ARE MICROSOFT. RESISTANCE IS FUTILE
YOU WILL BE ASSIMILATED

STOP THE VIOLIN
VISUALIZE WHIRLED PEA

MY COMPUTER DOESN'T UNDERSTAND M

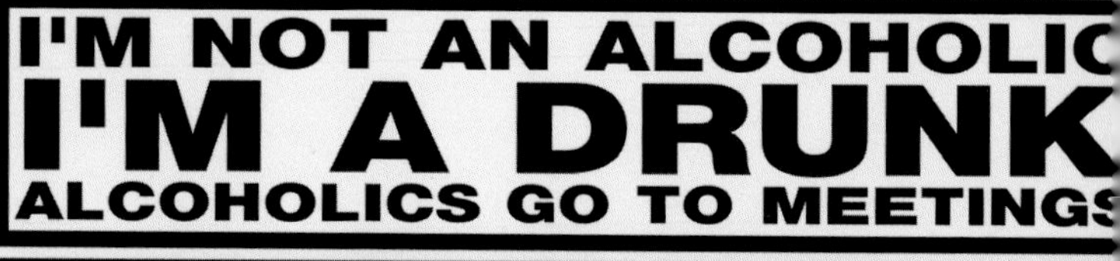

I'M NOT AN ALCOHOLIC
I'M A DRUNK
ALCOHOLICS GO TO MEETINGS

SOMETIMES I WISH LIFE HAD SUBTITLE

URINE NATION

A question here: Didn't piss, much less the golden shower implied by Calvin's stream raining down on one's ex, used to be more transgressive? In the past decade, we've seen Helen Chadwick's *Piss Flowers*, Tony Tasset's *I Peed On Myself* , Knut Åsdam's *Untitled: Pissing* and even Keith Boadwee's *Untitled* enema paintings. But none of them had the same impact as Robert Mapplethorpe's golden shower boys *Jim and Tom, Sausalito* (1977) and Andres Serrano's ubiquitous *Piss Christ* (1989), which between them virtually killed the National Endowment for the Arts. When did piss become shtick? Much of this slide from the obscene to the merely bawdy can be seen as a result of the gargantuan flood of pornography on the net. What is interesting about this deluge (just can't give up these fluid metaphors) is that certain kinds of established practices withered

THE SORT OF FETISHES THAT HAD ONCE BEEN THOUGHT TO BE LIMITED TO GAY DEMIMONDES AND EUROPEAN DECADENTS BECAME AVAILABLE ALL DAY EVERY DAY IN PEORIA TO ANYONE WITH A MODEM, A MONITOR, AND A SEMBLANCE OF PRIVACY.

MASS MARKET MAGAZINES

featuring hardcore Euro-kink were unimaginable a decade ago, but it's now de rigueur in PENTHOUSE and HUSTLER.

IN THE DARWINIAN WORLD OF PORN, *COMPETITION HAS TO BE MET AND BESTED OR THE VENUE WILL DIE.*

Once they've seen Paris, how ya gonna keep 'em down on the farm?

URINE NATION

Vhen average pornsters get used to Web sites 'eating fisting and pig fucking, how do you 'eel them back to the newsstand? I'm talking about those guys who wear their hair short on the top and long in the back (the ShoLo or Mullet), speak fluent Pissing Calvin, and spend their leisure time flipping between tapes of backyard wrestling, watching porn, and listen-ng to nü-metal. Magazines had to offer at east a part of what everyone was already get-ting online, and often for free, if they expected to maintain any market share at all. And what happens in the soggy pages of skin mags, especially in combination with the pornotopia of the net, has a huge effect on what the rest of the culture considers transgressive.

This is how we arrive at the point when mass culture not only accommodates but regularly deploys representations of practices formerly confined to Sade's *120 Days of Sodom* and Bataille's *Story of the Eye.* Thus we have ads for Candies shoes featuring ex-Playmate Jenny McCarthy sitting on the can with her panties down at her ankles, co-ed bathrooms complete with flushing sounds on TV's *Ally McBeal*, and huge bus ads for the Adam Sandler movie, *Big Daddy*, featuring the star and his diminutive ward relieving themselves on the side of a wall.

I invoke the present banality of piss not to bolster claims of the decline and fall of the Republic, but rather to contextualize the pornographic within the aesthetic economy of information technology. If you accepted the notion of computers as "solitude enhancement machines" that I advanced earlier, perhaps you'll follow as I go even farther, invoking a new law, to join Moore's on the price and power of processing, and Metcalfe's on the additive value of networks.

URINE NATION

MY CONTRIBUTION:
"Early adopters are chronic male masturbators." Like Moore and Metcalfe, I have no final proof for my law, but I ask you: Who bought the first VCRs from 1975-1982? Were participants in the earliest AOL chat rooms logging on at ruinous per-minute charges to talk about recipes?

What are business travelers in anonymous hotel rooms watching on all those DVD-equipped laptops? Not to be sexist here, but in the first two years after the service was generally available, was anyone you knew who installed a high-speed Internet connection to their homes a single woman?

We're still quite reticent about sex media's place within digital culture and the so-called new economy, but even telecom executives will admit to the *New York Times* that porn is "the crazy aunt in the attic. Everyone knows she's there, but you can't say anything about it." I'd say it was the lonely uncle with a cable modem and Pissing Calvin on his truck, but I agree in spirit.

GROWING UP PULP

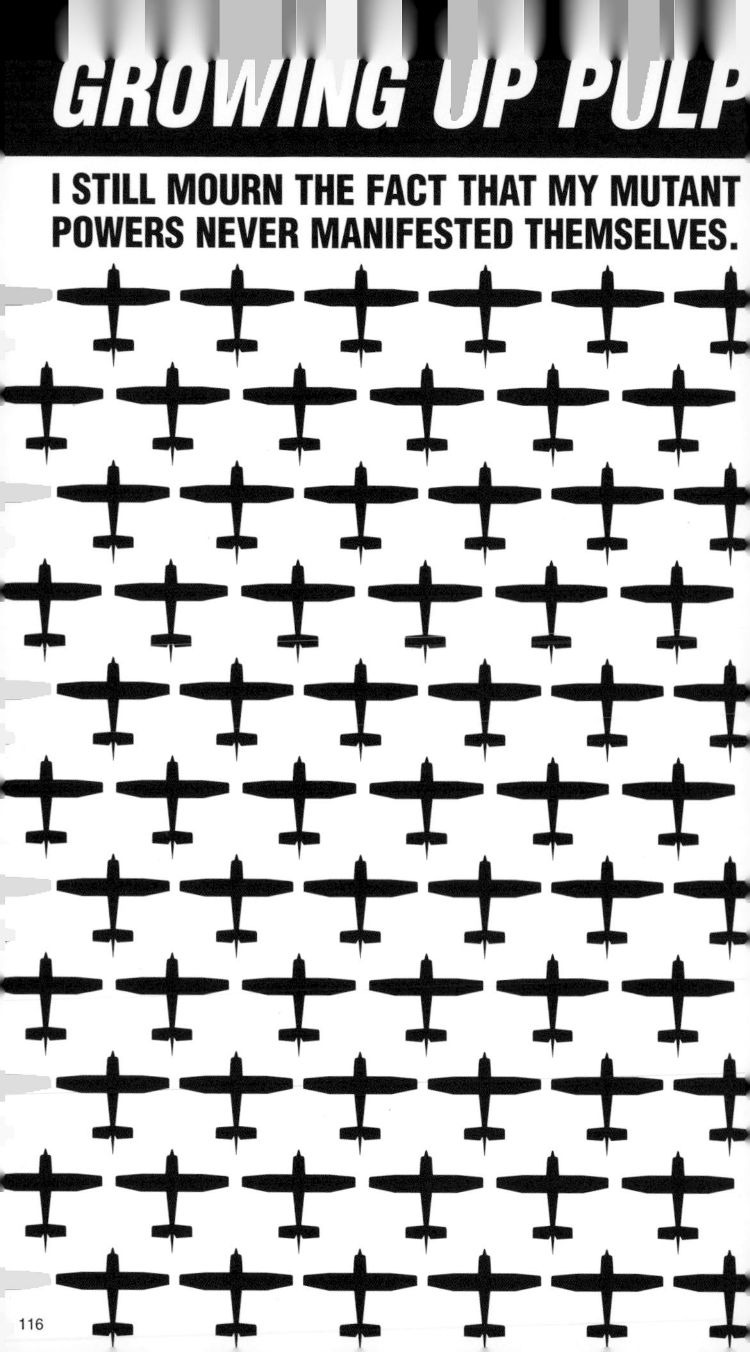
GROWING UP PULP
I STILL MOURN THE FACT THAT MY MUTANT POWERS NEVER MANIFESTED THEMSELVES.

GROWING UP PULP

This is a neurosis to which I was doomed by the singular figures of Marvel Comics publisher Stan Lee and his under-appreciated collaborator, Jack "King" Kirby. Lee, who wrote so many comic books that he claims to be the most prolific storyteller in human history, had an imagination that was almost perfectly reflective. The world worried about nuclear power, so he bombarded his characters with Gamma rays (creating the Hulk), or just had them bitten by radioactive insects (Spiderman, of course). In the early '60s, Americans believed in the domino theory, so he created Iron Man, a millionaire inventor of the Howard Hughes variety who donned transistorized armor to battle Vietnamese commies and a Russian doppelgänger called the Crimson Dynamo.

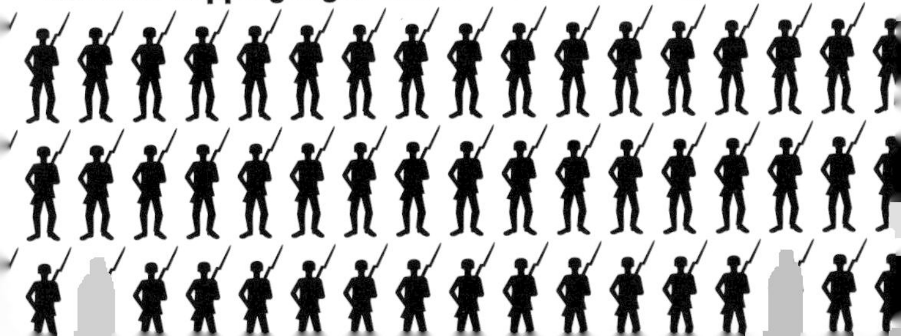

By 1963, Lee and Kirby came up with one of the most enduring of their creations:

THE X-MEN

The X-Men were, in fact, not men at all, but rather teenagers, and ones who began to exhibit "strangeness" just as they hit puberty.

GROWING UP PULP
THE BEAST BEGAN TO LOOK MORE AND MORE APE-LIKE
THE ANGEL SPROUTED WINGS,
AND CYCLOPS COULD ONLY CONTROL THE DESTRUCTIVE BEAMS SHOOTING OUT OF HIS EYES BY WEARING A SPECIALLY CONSTRUCTED VISOR.

GROWING UP PULP

Here was a metaphoric reading of adolescence worthy of, and perhaps besting, Freud's. Trumping the lumpy, hairy and sebaceous embarrassments that usher most boys into secondary-sex characteristics, the X-Men were a beacon of hope through the long, dark nights of an eighth grader's soul.

HOWEVER, AS YOU MAY HAVE INTUITED, I HAVE YET TO TURN INTO AN INVINCIBLE, TELEKINETIC CHAMPION OF JUSTICE.

GROWING UP PULP

While growing up pulp left me with that ambition as yet unfulfilled, at least it shaped my taste, especially the way I came to aestheticize technology. The pulp imagery aimed at boys has always been filled with all manner of gimcracks, machines, and–above all–the gleaming, homoerotic fetishism of weaponry. Science fiction literature and the fantastic cinema both offer a kind of adolescent sublime, and comic books do too, but as a medium, comics add a complex layer of schematics as well. Their crude, four-color pages open onto a model-ready futurism: cutaways of Professor Xavier's School for Gifted Youngsters (revealing the X-Men's battle training rooms), of the Bat Cave's supercomputer beneath stately Wayne Manor, and–most complex of all–the Fantastic Four's headquarters in the Baxter Building. The Baxter Building was supposed to be in midtown Manhattan, just forty imaginary blocks from the very spot in which I was raptly reading about it, and had been modified by Reed Richards (the rubber-limbed scientific genius known as Mr. Fantastic) to be the most connected, tech-filled, helipad-equipped building the world had or would ever see. In other words, this was a pulp imaginary that prepped me for the screen-blanketed, information-overloaded, communication-crazed networked environment I live in now–only without the tights, and those villains from the Negative Zone.

GROWING UP PULP

For the first time since their original run in the 1960s, Jim Steranko's *Nick Fury: Agent of Shield* was recently put together as a bound collection. In this subterranean yet hugely influential series, Steranko brought a sense of the high modern moment to '60s comics. He was one of the first to incorporate photographic imagery directly into comic book layouts. The stories themselves were neither any more nor any less histrionic than the rest of the Marvel line at the time, influenced though they were by James Bond, the *Man from U.N.C.L.E,* and the rest of Cold Warrior spydom. My favorite moment in the series comes when Fury, the hard-bitten hero, returns to New York from his orbiting Helicarrier for some well-deserved R&R. He steps into an apartment-cum-Xanadu that today would make the readers of *Wallpaper** either wet or hard, depending on their fluid preference. It's like the set design from Joseph Losey's Liz Taylor and Richard Burton mod fiasco *Boom!* or Julius Shulman's photographs of Richard Neutra's Case Study House, but its insertion into the ever more surreal surroundings of a 12¢ comic book makes it multiply mind-blowing.

GROWING UP PULP

In retrospect, I've been thinking about how this pulp modernism occasionally intersected with the high modernism of the '60s and the postmodernism of the '70s. Pop that pops to mind includes Roy Lichtenstein's comic panel paintings, Archigram's walking cities, and Öyvind Fahlström's *Meatball Curtain (for R. Crumb)*. When I first saw Claes Oldenburg's *Lipstick (Ascending) on Caterpillar Tracks* as a kid, it didn't bring Duchampian readymades to mind so much as Superman's enormous key to the Fortress of Solitude, an object so big that only the last son of Krypton could fly it into the lock. There are subtler connections as well, like Robert Smithson's original magazine layouts from the '60s and '70s of essays like "Quasi-Infinities and the Waning of Space" and "Strata: a Geophotographic Fiction." Their particular form would have been unimaginable without the comic paradigm to guide him.

GROWING UP PULP

Those of us who grew up pulp were softened up for expensive watches with lots of gimmicky features, for handheld communicators and Personal Data Assistants, for Wi-Fi everywhere, for smart cars, and the sorts of toys that used to be so popular in the *Sharper Image* catalogue. You could craft an investment strategy around the pulp fictions of the decade before, the formative influence on the generation of inventors, investors and consumers at work today.

GROWING UP PULP

Just as pop music is tied inexplicably to the sexual imagination—that's why no music ever has the same poignancy or power over us as that we listen to during our first sexual awakening—so, too, the pulp-tech of our youth makes an indelible impression on our scientific and engineering paradigms. If you look back ten or fifteen years, pulp took a distinct turn away from the mechanical and toward the biological. The layouts and art became more fluid, the organic dominating the electronic. What that means for the stock market I can't say, but maybe it holds out hope that I'll finally get my powers.

EXTRUSION
VERTIGO

EXTRUSION VERTIGO

I live in Los Angeles and suffer from a native malady: extrusion vertigo. When bit players from TV shows work out at my gym, Sylvester Stallone drives by, I do a studio visit with Farrah Fawcett (turns out she's a sculptress/actress), or other such flat screen characters manifest themselves to me in three-dimensional form, I get dizzy.

My perceptual apparatus ties itself into knots trying to flatten the celebrity in question into his or her recognizable configuration while at the same time leaving the environment intact. I need a second opinion, but my uncharacteristic (at least for L.A.) hostility to celebrity-driven spectacle culture may at root simply be a corollary symptom of my condition.

EXTRUSION VERTIGO

WHY ELSE WOULD I STILL BE SO ILL-TEMPERED ABOUT OUR CULTURE'S OBSESSION WITH STARS?

You'd think that my neighbors in the heart of the image factory would be as annoyed with extrusion vertigo as I am.
But that would underestimate the fervor with which everyone in Hollywood feeds off the perpetual dream machine at its center.

One would naturally assume that the ***INDUSTRY GRIPS, PERSONAL ASSISTANTS, SECOND-UNIT DIRECTORS AND EFFECTS PRODUCERS*** would grow tired of the stars who inhabit their own, even more tightly wound cocoon, but in fact, the opposite is true.

The below-the-line toilers are fans first and foremost–that's why they fought their way into showbiz–and the flame burns brighter in Bel Air, Beverly Hills and Brentwood than it does in far off Baltimore, Brussels and Beirut.

EXTRUSION VERTIGO

IT WAS A COMMONPLACE OF THE OLD NEW ECONOMY THAT AS COMMUNICATION NODES ARE DRAWN INTO EVER TIGHTER CONNECTIONS WITH EVEN BROADER BRANDWIDTHS, OWNERSHIP OF THOSE NODES SHOULD BE CONCENTRATED INTO EVER MORE BULBOUS STRINGS OF CONGLOMERATED NAMES. DURING THE TECH BUBBLE, IT WAS BARELY SATIRE TO IMAGINE AN ADVERTISING CAMPAIGN ALONG THE LINES

OF:

EXTRUSION VERTIGO
ONE
WORLD
ONE
MARKET
ONE
URL
<www.TimeWarnerDisneyClearChannelBertelsmanViacomTribuneNewsCorp.com>

EXTRUSION VERTIGO

In a "synergistic" media ecology like ours, the pressure to develop, sustain, and recycle celebrity is tectonic. We have to remember that we're far beyond the primitive stage of defining specific media – movies, television, records, magazines and the like. We now live in a Klein bottle of tesseracted media commodities: lunch boxes are imprinted with URLs pointing to Web sites promoting films by offering trading cards spun off from video games which link to personal fragrance products mentioned on Christmas albums that come packaged with commemorative T-shirts. I lost my innocence about this years ago in a Turkish bazaar when I looked up to see Tom Cruise's face, circa *Top Gun*, staring down at me, woven into a small rug. Neither the Islamic prohibition against idolatry nor my own naiveté about the "exotic" could stand up to Entertaindom's global reach.

I FLASHED BACK TO THAT RUG IN THE MIDDLE OF LISTENING TO MARIANNE FAITHFULL IN CONCERT, OFFERING A SMOKY RENDITION OF BLIND WILLIE JOHNSON'S "JOHN THE REVELATOR," A SONG OF ALMOST INFINITE STRANGENESS, MYSTERY AND POWER. FAITHFULL, WHOSE RASPY VOICE AND WORLD-WEARINESS MAKE HER SEXIER THAN A THOUSAND TEEN QUEENS, WAS SINGING HER VERSION OF A RECORD PRESSED IN THE LATE 1920S BY JOHNSON, WHOSE BIOGRAPHY OFFERS EVEN MORE HORRORS THAN USUAL FOR A DELTA BLUES-MAN. BLINDED BY LYE AT 7, JOHNSON DIED OF PNEUMONIA AT 41 WHEN A WHITES-ONLY HOSPITAL WOULDN'T ADMIT HIM AFTER HE WAS FORCED INTO A TORRENTIAL RAINSTORM BECAUSE HIS HOUSE BURNED DOWN TO THE GROUND (THE REST OF HIS LIFE WASN'T MUCH CHEERIER). "JOHN THE REVELATOR" CAME TO PROMINENCE WHEN IT WAS INCLUDED IN THE *ANTHOLOGY OF AMERICAN FOLK MUSIC* 25 YEARS LATER.

EXTRUSION VERTIGO

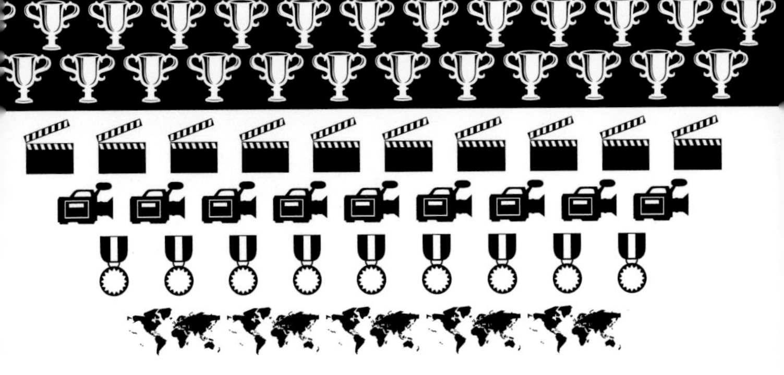

EXTRUSION VERTIGO

The *Anthology* is the first great bootleg, a compilation of commercially released records collected by Harry Smith, a seminal avant-garde filmmaker, anthropologist, self-proclaimed alchemist, and inveterate scavenger of objects ranging from Ukrainian Easter eggs to paper airplanes to pre-WWII 78 discs. The *Anthology* has been credited as the source of the '50s and '60s folk revival, and through its influence on those folkies who went electric (think Bob Dylan and the Band), with being a Rosetta stone for rock 'n' roll as well. Smith's *Anthology* was not simply collected, it was curated: a dark sonic portrait of a primal America which generations of artists have felt impelled to absorb and rework. People who find each other through the *Anthology* speak fondly of meeting in Smithville, not so much a place as a mode of being.

Faithfull took part in the Harry Smith Project, a concert/happening that originated in London, migrated to New York, and had its third installment in L.A. Other performers included Elvis Costello and Beck, and this time, the Project followed a film screening, folk concert, and groundbreaking interdisciplinary symposium sponsored by the Getty Research Institute. At the concert I enjoyed seeing myself as a participant in a recursive technological loop. The songs tend to predate recording technologies (they were first released on 78s); the advent of the LP allowed Smith to achieve his curatorial objective in a longer format; the CD reissue of the *Anthology* by the Smithsonian sparked yet another return to Smithville; after the reissue you could find these tunes ripped from their CDs and available over the net on file sharing services; and there I was, back in a community of listeners, hearing them live. So, why in the midst of this celebration was I reminiscing about that Turkish bazaar? I think it was the disjunction between this Getty-sanctioned apparatus of pop stars and academic experts and the man (Smith, not Cruise) himself.

EXTRUSION VERTIGO

Smith lived a kind of pure bohemianism, he and his peers constituting an avant-garde that simply doesn't exist anymore. Smith was a mythmaker of the first order, and one hell of a user of other people, but money and general renown really did not seem to matter much to him. Although he died little more than a decade ago–frail, indigent and (where else?) in the Chelsea Hotel – it seems harder and harder to imagine creativity so unconnected to capital. By coming to Smithville, Faithfull, Costello, Beck, and all the rest were further complicating my extrusion vertigo. Here's a confession: these are the celebrities I like. I buy their music without troubling myself as to whether or not they "sold out" by pledging their considerable talents so fully to the pop culture machine. And yet the Tom Cruise rug disrupted my rapture. It's all well and good to announce the end of the avant-garde and the apotheosis of design and entertainment, but this realignment creates a mysterious void at the very center of our culture now. The cartoonist Chris Ware puts it well: "Sometimes I really wonder about the efficacy of any art at all, but that's probably because we live such privileged, genuinely rich lives as Americans. If I were imprisoned for any amount of time, I'm sure it would all become much clearer." I keep thinking that somewhere in Smithville, there's an answer to all of this that doesn't entail incarceration; that we just have **TO KEEP LOOKING.**

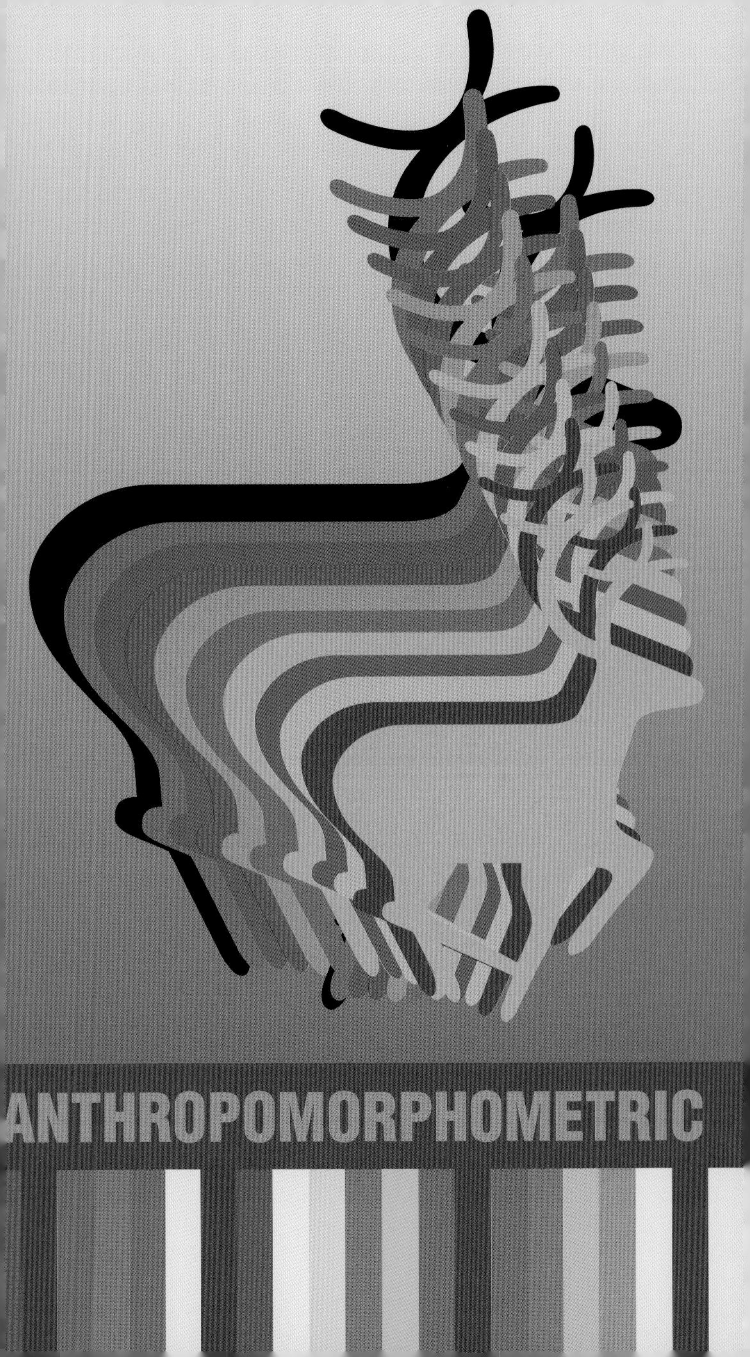
ANTHROPOMORPHOMETRIC

ANTHROPOMORPHOMETRIC

THE HUMAN ANIMAL
MAY BE THE ONLY ONE TO DERIVE PROFOUND,
indeed visceral pleasure from the contemplation of its own image

(YES, I'M TALKING ABOUT YOU)

The peacock and the butterfly are beautiful, of course, but they dress for dating and mating success, not to gaze longingly at their own mug reflected in the mirror.

So, if narcissism – like tool making and fetishism – is one of those definitive human traits, it's worth thinking about the extent to which self-love has likewise shaped our built world.

ANTHROPOMORPHOMETRIC

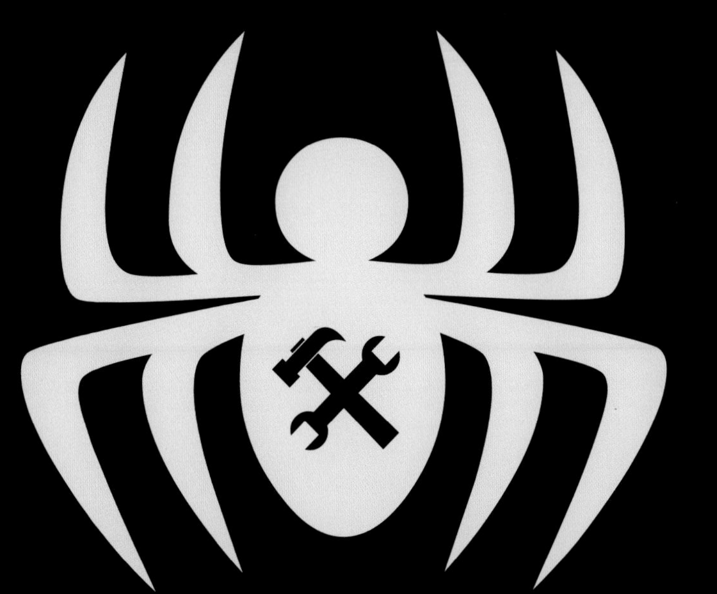

ANTHROPOMORPHOMETRIC

We are hard-wired to search out the human form, and we discover it everywhere. We hear the "whistling" of the wind and the "wail" of a siren. We see the man in the moon. We scan cars' grills when we hit the dealer's lot, and expect auto designers to take it as axiomatic that Pontiacs and Jaguars will have countenances as distinct as their engines. Watches, toasters, and even some personal digital data assistants are designed to mirror the proportions and symmetry of our own smiling faces. O brave new world, that has such simulations staring back at us, unabashed

There's a range of relationships between narcissism and anthropomorphism, of course, and also with what interface designers refer to as "personalization."

When we see two headlights and a bumper as a face, we are anthropomorphizing; but when we ascribe personality to those cars ("she's cranky whenever it gets cold"), that's personalization.

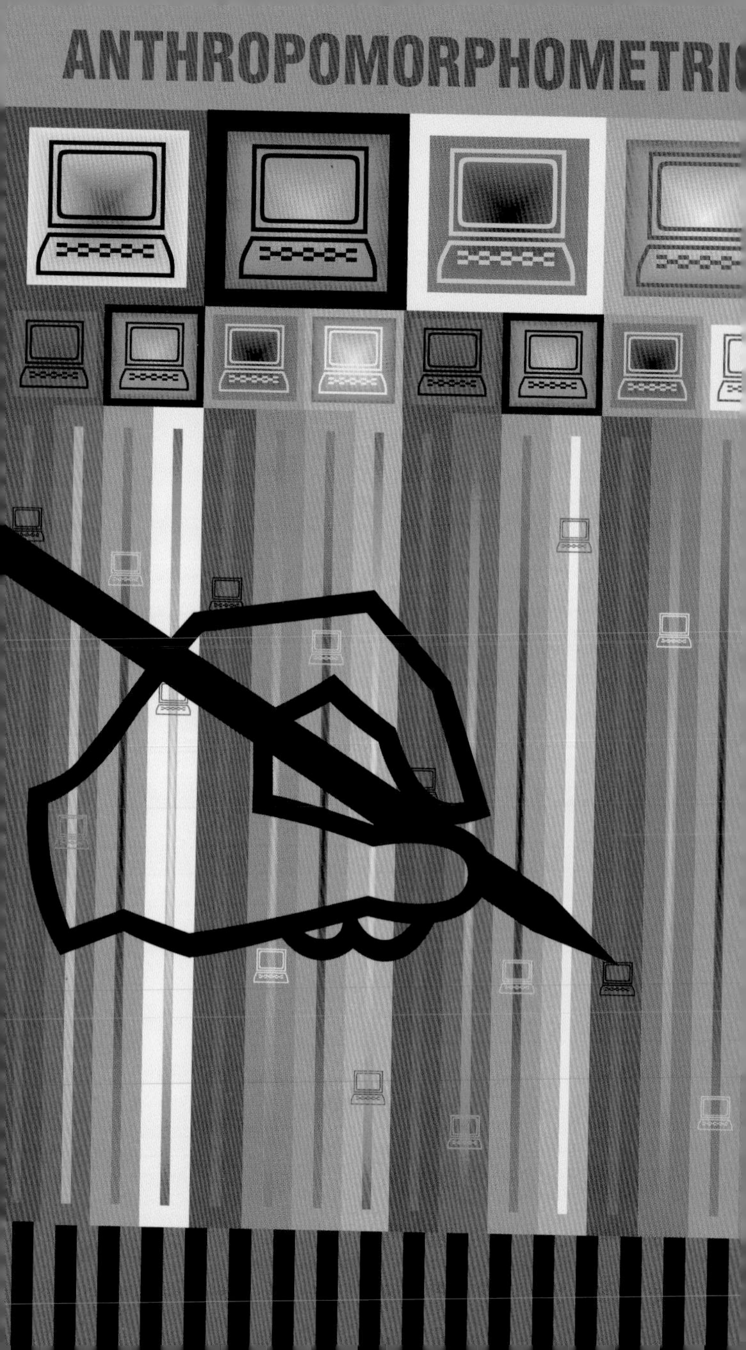
ANTHROPOMORPHOMETRI

ANTHROPOMORPHOMETRIC

WHILE WE KNOW, AT LEAST OBJECTIVELY, THAT PERSONALIZING OBJECTS AND TECHNOLOGIES DOESN'T ACTUALLY MAKE THEM CONSCIOUS, THE PHENOMENON ENDURES.

Thus the staying power of the ancient hacker joke:,

DON'T *ANTHROPOMORPHIZE* COMPUTERS; THEY HATE THAT!

In interface design, one of the raging debates over the past few years has been just how much digital systems should, in fact, mimic human behavior and social interactions. As processing and simulation technologies improve, systems are better and better able to abandon pop-up boxes marked simply, "Address," for example, and to use natural language instead, in which the interface prints or says, "Beatrice, could you do me a favor and type in your address in for me?"

THE CULTURAL IMAGINARY OF SCIENCE FICTION BOTH PREPARES US FOR THIS STATE OF AFFAIRS AND ENSURES A CERTAIN DISAPPOINTMENT THAT SEGUES EASILY INTO DISTASTE.

That is, the talking computers of *Star Trek* promise that we, too, will be able to chat with our smart machines, but the absurdity of our laptops' mechanical rhetoric falls so far short of the mark that the experience is actually creepy. Even when it works, it doesn't. A remarkable telephone interface called Wildfire came out a few years ago. The perky female voice of the digital receptionist was technologically impressive–and profoundly *unheimlich*. When "she" stuttered or hemmed and hawed, it was less a simulation of the real than a *tableau mordant*–a reminder of the vast, perhaps unbridgeable gulf between the "anthro" and the "morphic."

ANTHROPOMORPHOMETRIC

Anthropometrics is a sub-specialty of pediatric medicine, and is devoted to mapping human growth, generating endless charts of "proper" age/height/weight.

What might be interesting to contemplate is how we could develop a way to measure our collective desire to remold the world in our own image; in other words, an anthropomorphometric.

TAKE MICROSOFT, FOR EXAMPLE, WHICH HAS A HISTORY OF CREATING INCREDIBLY AWFUL INTERFACES. THEIR FORAYS INTO HUMANOID HELPMATES HAVE BEEN EVEN WORSE. AN ANTHROPOMORPHOMETRIC COULD HELP US EVALUATE THE WORD PROCESSING SOFTWARE PACKAGE WHICH I USED TO WRITE THIS ESSAY.

It features "Max" the Office Assistant, a dancing little Mac Classic with legs to "help" users navigate through the bloated word-processing software. This little creature appears in certain situations to offer assistance. If you type a date, skip a line or two and write, "Dear John," followed by a comma, up pops our little friend with a glowing light bulb displayed on his (?) screen and a floppy disc port that curves into a wan and unconvincing smile: "It looks like you're writing a letter. Can I help?" When you flail desperately at the keyboard and finally make it\"him" go away, he wiggles through a pathetic goodbye dance, or a hand appears on the screen and gives a weak little wave like the least popular kid in the tenth grade, his horribleness only exacerbated by this last desperate gesture to tug at your attention. As bad as this is, Microsoft's abandoned interface character "Bob" from 1995 was even worse, and in an equivalent to the Kelvin scale, could be rated as absolute zero on the anthropomorphometric scale. As other pixellated humanoid icons and three-dimensional avatars appear on our screens, at least we'll be able to judge them with some sort of consistency.

An anthropomorphometric could help us in other areas as well. Think about the question of authorship. Though arguments about the death of the author now seem long dead themselves, the development of open-source cultural initiatives and communal creation on networks is breathing new life into the question of how to create what might be called "proper name"-brand awareness.

THERE MAY WELL BE A RISING ANTHROPOMORPHOMETRIC IN ARCHITECTURE AND DESIGN, FOR EXAMPLE, WITH THE EMERGENCE OF A STAR SYSTEM, SIGNATURE STYLES AND "AUTHORED" BUILDINGS AND OBJECTS BEING ALL THE RAGE.
(Painting went through this centuries ago, with the rise of the artist and the end of the studio.)

<u>*ARCHITECTS*</u> spent much of the 20th century turning their names into signatures, and their very personas into modes of access to the new.

Think of how Richard Meier adopted Philip Johnson's adoption of Le Corbusier's black, round spectacles to nail the look of the modernist architect (pity Meier couldn't cadge the master's talent, too).

In the 1910s, certain actors realized that people were coming to see them, not just the movies, and demanded to be named, and paid accordingly. After a few battles, the producers caved in, if only because they saw it was in their best interests to have these stars identified as more than simply the "Biograph Girl," as Florence Lawrence, the first real movie star, had been known.

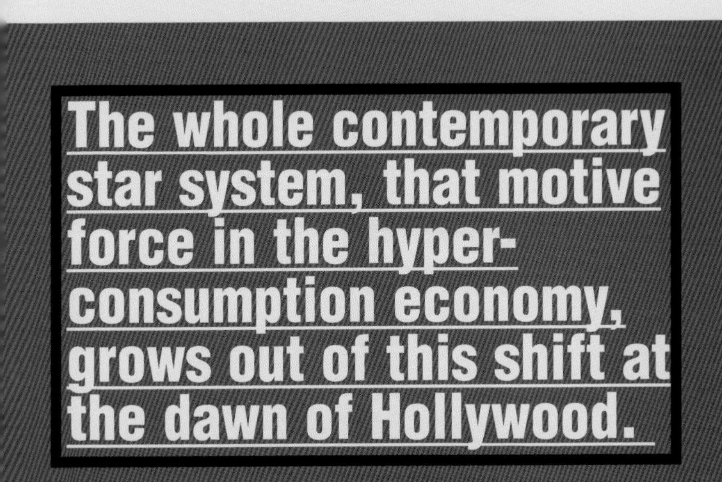
The whole contemporary star system, that motive force in the hyper-consumption economy, grows out of this shift at the dawn of Hollywood.

There was another shift in Hollywood sixty years later with marketing strategies developed around the director's name. While the French had been taking about a *politique des auteurs* for two decades, it was only in the 1970s – the great era of Altman, Scorsese, Coppola, etc. – that the studios realized that they could create a larger market by anthropomorphizing the formerly anonymous figure of the director for the film-going audience. What had once been Hitchcock's and Capra's alone has now spread to the horrible "A film by..." credit to first-time directors of truly limited talents.

ANTHROPOMORPHOMETRIC

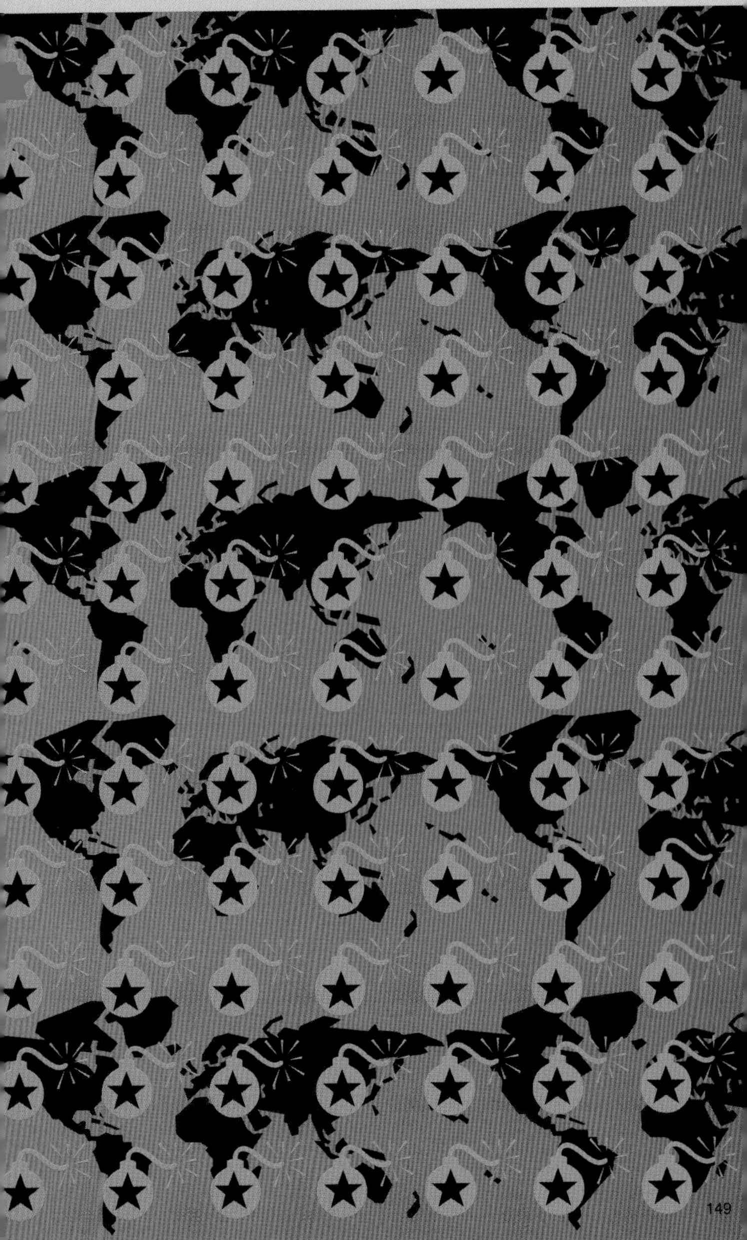

ANTHROPOMORPHOMETRIC

With a product designer like Karim Rashid getting a New Yorker profile and Bruce Mau's name dropped only slightly less often than Bruce Weber's, can designers be far behind? This naming and concurrent celebrity is an inversion of narcissism: we look for a (partial) reflection of ourselves in the (large-scale) mirror stage of the other. The next (one can never say final) step in this is the development of the celebrity DJ and the signed mix. DJs archive, retrieve, and deploy sound, so we might liken their burgeoning renown as a measure of the anthropomorphometric of the database itself.

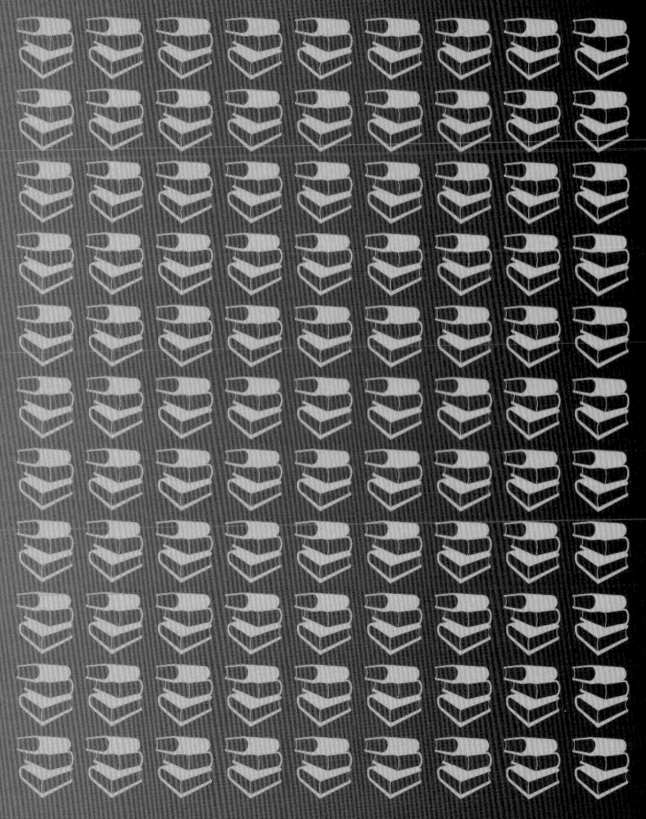

FIGURE/
GROUND

FIGURE/GROUND

Gestalt psychology never did much for me, but those figure/ground experiments were pretty cool. You know the classics: a drawing that can be either a vase or two faces in profile; the rabbit that's a duck that's a rabbit; and my adolescent favorite, the portrait of Sigmund Freud that's also a naked woman. When you look at an image, the figure is supposed to have the definite shape, the prominent contour, and, to use the peculiar, Germanic phrasing beloved of Gestaltists, a greater "thing character" than the ground. In the best of the figure/ground illustrations, there's a moment when your perception "pops" and what had been the ground flips instantaneously into the figure: it becomes "thingy." This is a transformation simultaneously magical and quotidian, and after the pop, it seems impossible that you ever didn't see the figure in precisely that way.

FIGURE/GROUND

FIGURE/GROUND

I'd contend that this kind of pop happens every once in a while in the culture at large: some "thing" emerges from the ground, while formerly prominent figures sink back in to the amorphous periphery. It's all there in *They Live*, an odd sci-fi film that John Carpenter directed at the tail end of the Reagan years. Carpenter's erstwhile hero, played by professional wrestler Rowdy Roddy Piper, is a drifter who finds a pair of glasses that allows him to see that the snooty, BMW-driving, Lacoste-wearing, double-espresso-sipping young urban professionals surrounding him were in fact creatures from another planet intent on sucking the earth dry. What weird brilliance Carpenter (through Piper) had lay in his ability to articulate that which was already present: Yuppies were indeed aliens, and their spawn at Enron and Worldcom sure enough hoovered every cent of venture capital out of our economy.

Rowdy Roddy Piper finesses the pop. This phenomenon is distinct from the paradigm shift, which (think Einstein and relativity) is essentially a narrative about invention. The paradigm shift ruptures the knowledge universe, whereas the figure/ground shift is fundamentally a transposition–the recognition of things that were already in the culture, but not central to its perception of itself. The pop between figure and ground is as recent as Malcolm Gladwell's notion of the "tipping point" and as ancient as the story of the Emperor's new clothes. The pop makes reference to a conspiracy theory so compelling it becomes history.

FIGURE/GROUND

One of the loudest pops of the last hundred years occurred as a result of the public-ation of Freud's *Interpretation of Dreams*.

WITHIN A GENERATION, WE WERE NO LONGER UNCONSCIOUS OF THE UNCONSCIOUS.

It's not that artists before Freud were oblivious to inexplicable motivations and hidden meanings (Macbeth's damned spot and all), but rather than after Freud, those factors which formed the ground for artists from Aeschylus to Shakespeare to Coubert were transformed into the very figure of work.

UNCOVERING THE UNCONSCIOUS IN POST-FREUD ART AND CULTURE WAS EASIER THAN FISHING WITH DYNAMITE. JUST THINK OF THE WAY POOR JIMMY STEWART WAS TREATED BY HITCHCOCK IN FILM AFTER FILM: CRIPPLED, CASTRATED, FRUSTRATED, AND GENERALLY FUCKED WITH.

IF YOU GO LOOKING FOR THE UNCONSCIOUS IN *REAR WINDOW, ROPE* AND *VERTIGO,* HOW THE HELL COULD YOU NOT FIND IT?

FIGURE/GROUND

FIGURE/GROUND

After the pop, the new relationship takes on the mantle of "common sense," of natural perception and transcendent truth. But it takes time. We can understand how early in the last century a woman of "refined taste" could still want antimacassars on her plush Victorian settee. But the conspiracy theory of modernism was just too good. The Cubists, Futurists, Purists, Constructivists and innumerable other avant-gardists flipped her world, wherein the industrial was the ground on which the figures of culture were so ornately drawn. The incendiary industrial chic of Duchamp's fountain, the anti-humanism of Le Corbusier's machines for living, the scopic mechanization of Vertov's "Kinoeye," Marinetti's wholesale embrace of speeding planes and motor cars: in these by now familiar icons, one was forced to confront the machine as the salient element of the 20th century visual culture, not simply its backdrop.

FIGURE/GROUND

If we employ the figure/ground flip as an aesthetic algorithm, we can input other signal moments as a way to map the development of our own common sense and transcendent truths. Sixties Pop flips the figures of high art and the ground of low culture; McLuhan flips the message and the medium; postmodern art/design takes the ground of historical reference and flips it into the figures of Sherrie Levine's re-photography of Walker Evans, Mark Tansey's historical tableaux of literary theorists Paul de Man and Jacques Derrida, Neville Brody's explicit reworkings of El Lissitzky's pioneering Constructivist layouts, and Paul D. Miller aka Dj Spooky that Subliminal Kid's audio/video remix of D.W. Griffith's *Birth of Nation*.

WHILE INDUSTRIAL MACHINES POPPED CLOSE TO A HUNDRED YEARS AGO, INFORMATION IS FINALLY EMERGING AS THE KEY FIGURE FOR THIS NEW CENTURY.

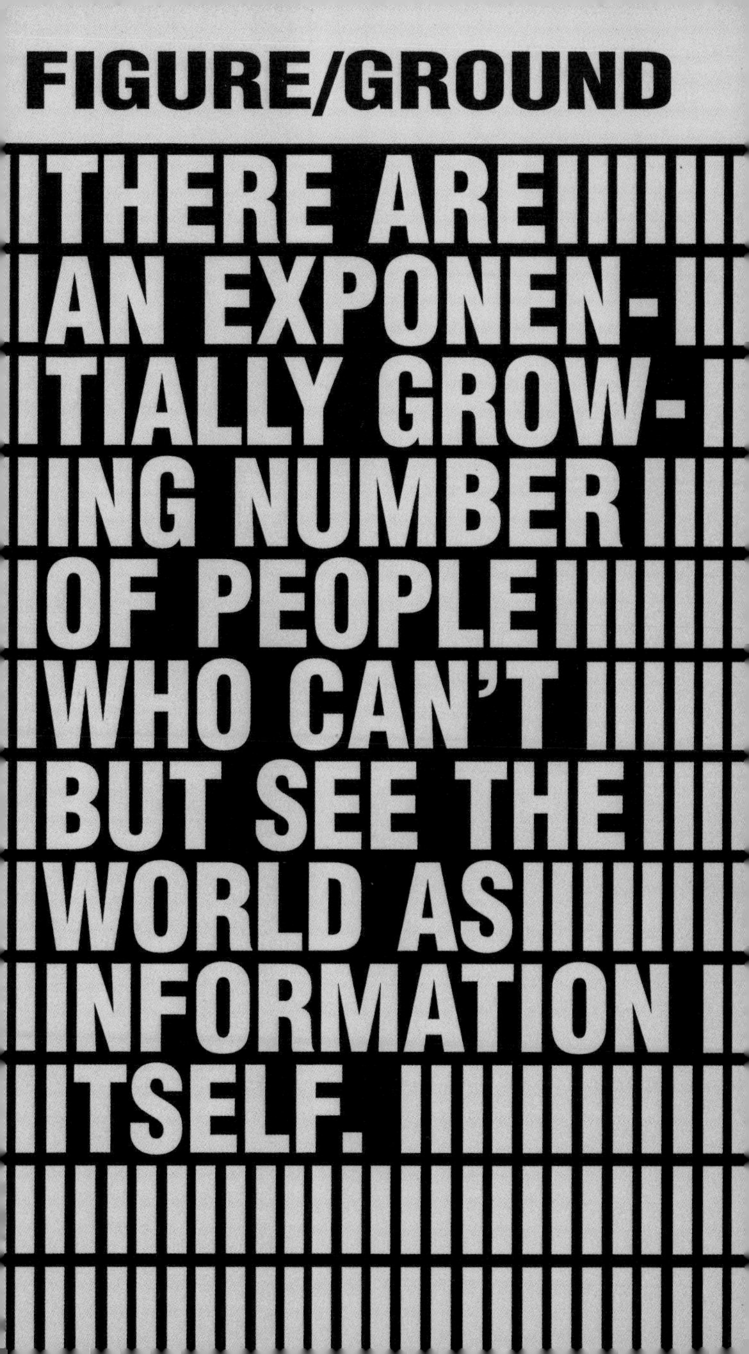
FIGURE/GROUND
THERE ARE
AN EXPONEN-
TIALLY GROW-
ING NUMBER
OF PEOPLE
WHO CAN'T
BUT SEE THE
WORLD AS
INFORMATION
ITSELF.

FIGURE/GROUND

Artist John F. Simon, Jr. rips apart the guts of a Powerbook to create a display space for the evolving software simulation of *ComplexCity*. Designers IOD craft their Webstalker software to give visual form to the sprawl of the network. Karen Kimmel and James Bond open KBOND, a men's store in Los Angeles, and base their entire sales strategy on the deployment of clothing as information accessory. Even Frank Gehry's Guggenheim Bilbao and Disney Concert Hall, those most sensuous of 21C signifiers, are seen as the manifestation of CATIA software. These artists, designers, programmers, and entrepreneurs engage in a never-ending process of tweaking their vision of the world, and at some point in that process their perceptions have flipped, and they have come to see the fact of data as being just as–or even more–important than data as fact. All this may explain why M.C. Escher, the poet laureate of figure/ground, is now and always has been the favorite artist of every hacker I've every known.

WIRELESS
COSMOPOLITANS
WIRELESS
COSMOPOLITANS

I never fully believe people when they profess their love for television, not as a guilty pleasure, but as an *amour* every bit as important to them as bibliomania or cinephilia. I'm tempted to delve deeper: "For real, you actually like that shit?" I'm restrained from this level of confrontation for fear of being branded an elitist (who am I to resist the tube's embrace?), a sexist (a domestic device, television has long been gendered female), even a homophobe (camp has morphed into one of broadcasting's dominant modes). But let's not be cowed by the specter of cultural studies, which treats watching TV as an innately subversive act, seeing every instance of pop cultural consumption as a site of resistance to the culture of consumption itself: this is post-'68 posturing at its weakest – easy theory, slack politics. Sheer loathing can be liberatory. My favorite take on the medium comes from the schizophrenic novelist Frederick Exley who, in the '60s, identified TV's own split personality: in its deference to our fantasies, TV sadistically reduces its victims to prattling infancy. But its embrace is tender; after no matter how long an absence, it welcomes us back to its "brown-nippled bosom."

In darkly optimistic moods, I hope that the future will look on our present 40-hour-a-week TV habit with the same incredulity with which we regard those eras when people spent 3 to 5 hours a day in prayer. But for now, the battle over television appears lost, 500 channels being 50 times less interesting than 10. Television is no longer just television, of course, as it has now swallowed pop music (via MTV), most of print (from the magazines that cover television's pseudo-events to the books written by its pseudo-stars), and, of course, the entirety of the cinema (VCRs, DVDs, that monstrous hybrid known as the home theater). The transnational corporations responsible have honed these media into an all-encompassing, multi-tiered, interlaced and entirely self-referential sphere of personal consumption built around the culture of the entertainment celebrity.

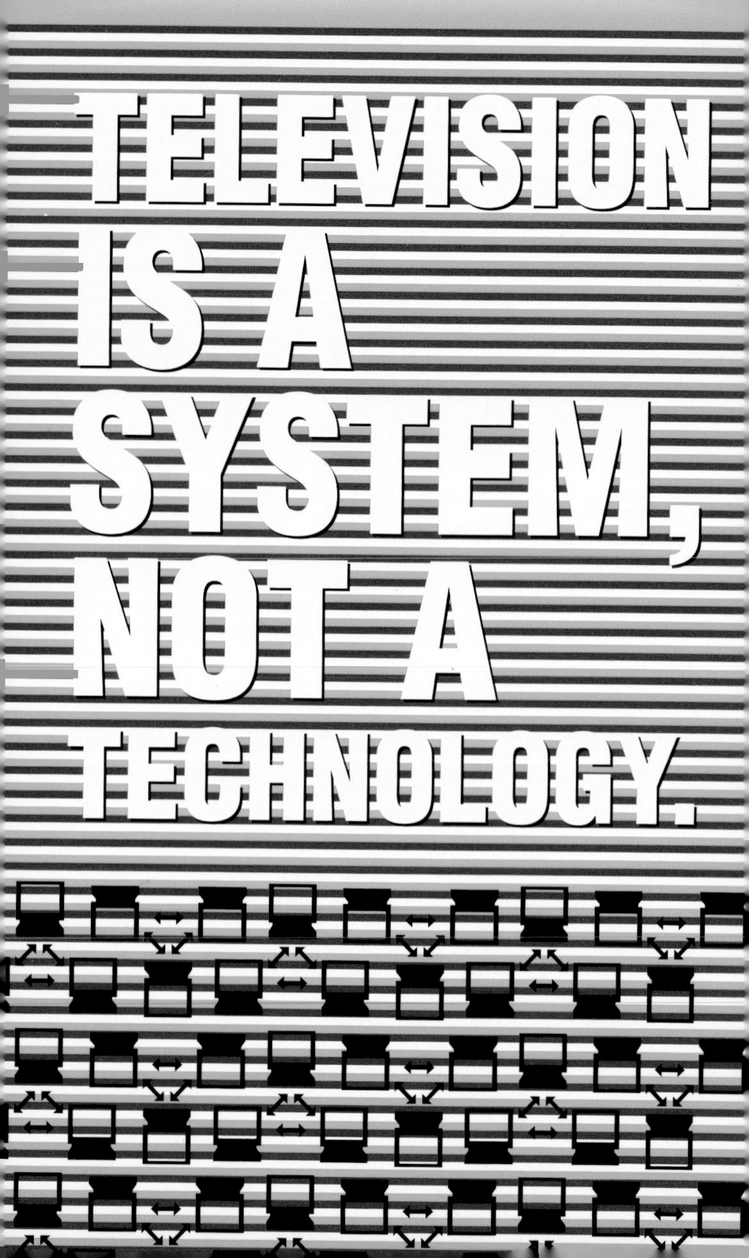
TELEVISION IS A SYSTEM, NOT A TECHNOLOGY.

THE INSURMOUNTABLE PROBLEM WITH TV IS THAT IT'S THE QUINTESSENTIALLY SUBURBAN MEDIUM; IT'S SMUG, INSULAR AND DOESN'T ASPIRE TO ANYTHING MORE THAN WHAT IT'S ALREADY PLEASED TO BE.

Descriptions of desolate housing tracts with living room windows emanating an eerie blue glow is a staple of European drive-by theories of America.

But take your car to the outskirts of Paris or Bangkok, and you'll realize that this is now the signature image of the whole world.

From the flappers and Dadaists to the hippies and punks, we've counted on youth culture to rupture bourgeois entertainments. As John Lydon of Public Image Limited once shouted,

"ANGER IS AN ENERGY."

What's happened in the past quarter century, though, is corporate subversion of adolescent enthusiasm through niche marketing: "anger is a demographic." How are viewers to escape the self-referentiality of their TV sets? To supplement anger, so easily commodified these days, they will need cunning, sophistication, and a community that does not inherit its culture so much as create it. In short, they will need to become cosmopolitan.

WIRELESS COSMOPOLITANS

Cosmopolitans strive to free themselves from the folk culture into which they were born or which they encounter upon arriving in a new locale (in this, they are anti-Romantic). They confound those who would build a demographic to engulf them, thus evading the spectacle's most banal manifestations. This is not to ignore the baggage– both comic and serious–that a word like cosmopolitan trails in its wake. Vodka-based cocktail; sex/lifestyle magazine for young women; the name above flophouses around the world: cosmopolitan is this and more. On regional television in Los Angeles during the 1950s, there was a daytime show based around the character of *The Continental*, a dulcet-toned European seducer. The show began with an anonymous female hand knocking at the door. The Continental, attired in a smoking jacket, opened it, gesturing to the viewer as though he could see her just as she could see him, suavely welcoming the audience of housewives to his sumptuous apartment, offering to light their cigarettes, complimenting them on their clothes and hair, and in general, creating a phantasmagoric direct-address simulation of sophistication and romance.

WIRELESS COSMOPOLITANS

When stolid American husbands grew to loathe the cosmopolitan competition *The Continental* created, they could draw from an ancient wellspring of suspicion. Ever since people began contrasting the term patriot with cosmopolite in the seventeenth century, there has been an undercurrent of hostility toward those who would proclaim themselves citizens of the world rather than members of the nation. By the late nineteenth century, the term "rootless cosmopolitan" came into use. Derided as unattached drifters, they were seen as lacking in allegiance to anything, disdainful of the simple pleasures and righteous outrage of the common folk, quick to take profit, and quicker to move on. Search the Web and you will see that the phrase endures in sites sponsored by the Liberty Lobby, the National Alliance, the Council of Conservative Citizens, and other race patriot sites, where–continuing a tradition now over a century old–"rootless cosmopolitan" is synonymous with Jew.

WIRELESS COSMOPOLITANS

It is not just the North American racist fringe that hates cosmopolitans; all over the world they are literally under attack. In the former Yugoslavia, hill dwellers claiming ancient tribal identities descended on urbanites in Sarajevo, whom they despised for their ethnic and religious tolerance. In Afghanistan, illiterate Taliban farm boys with surplus Kalishnikov rifles terrorized educated urban women, driving them out of the workplace, denying them access to schooling, and forced them back under the veil. Back home, less bloody, yet no less oppressive, voters in Missouri's and other "retro" states pass constitutional amendments to limit the freedoms of homosexual couples to marry in the "metro" areas of San Francisco and Boston.

Those who would produce and consume cosmopolitan media should not be surprised, then, to find they have enemies. But in the advanced-industrial-entertainment state that was once known as the West, enmity is not half the impediment that indifference can be. The computer will be central to cosmopolitan media, but digital technologies, even though they have spread around the world, still call Silicon Valley home. The Valley is an ill-defined, ever-expanding region south of San Francisco, a dispersed and undistinguished series of run-on suburbs which have grown to house the interlocking set of scientific engineering, and entrepreneurial communities that shape the future of technological media.

For the past few decades, perhaps driven by the anonymity of their work and home environments, the Valley's techno-elite have been promoting the wireless ideology, the notion that fleeing cities for the country, linking up by satellite phone and modem, and typing from the porch, the beach, the forest (choose your favorite idyll and insert here) is the pinnacle of personal freedom, as opposed to asocial detachment. If you ask me, the medieval motto, "city air makes free," proves that serfs knew better than Web surfers that freedom is more likely to be found in a crowd than home alone, staring into the screen.

USER
COLOPHON

BY
PETER LUNENFELD

VISUALS BY
MIEKE GERRITZEN

PRODUCTION BY
ALL MEDIA FOUNDATION AMSTERDAM (info@all-media.info)

PRINTED IN SINGAPORE

PUBLISHED BY
THE MIT PRESS
CAMBRIDGE, MASSACHUSETTS
LONDON, ENGLAND

AND

ALL MEDIA FOUNDATION
AMSTERDAM

MEDIAWORKBOOK
FIRST EDITION 2005

Peter Lunenfeld is the author of *Snap to Grid: A User's Guide to Digital Arts, Media & Cultures* (2000) editor of *The Digital Dialectic: New Essays on New Media* (1999) and editorial director of the Mediawork series (2001-2005), all from the MIT Press. He is a professor in the graduate Media Design Program at Art Center College of Design in Pasadena, CA, and director of the Institute for Technology and Aesthetics (ITA).

Mieke Gerritzen is founder and director of NL.Design, an Amsterdam based design company, permanently under construction. NL.Design makes designs for all media and works with many different designers, writers and artists. She is the head of the design department at the Sandberg Institute in Amsterdam. www.nl-design.net

All Media Foundation is an Amsterdam based foundation. All Media produces films, events like "The International Browserdays" and "Biggest Visual Power Shows" and books about media and design. www.all-media.info

SERUSERUSERUSERUSERUS
ERUSERUSERUSERUSERUSER
SERUSERUSERUSERUSERUS
ERUSERUSERUSERUSERUSER
SERUSERUSERUSERUSERUS
ERUSERUSERUSERUSERUSER
SERUSERUSERUSERUSERUS
ERUSERUSERUSERUSERUSER
SERUSERUSERUSERUSERUS
ERUSERUSERUSERUSERUSER
SERUSERUSERUSERUSERUS
ERUSERUSERUSERUSERUSER
SERUSERUSERUSERUSERUS
ERUSERUSERUSERUSERUSER
SERUSERUSERUSERUSERUS
ERUSERUSERUSERUSERUSER
SERUSERUSERUSERUSERUS
RUSERUSERUSERUSERUSER
SERUSERUSERUSERUSERUS
RUSERUSERUSERUSERUSER
SERUSERUSERUSERUSERUS
RUSERUSERUSERUSERUSER
SERUSERUSERUSERUSERUS
RUSERUSERUSERUSERUSER
SERUSERUSERUSERUSERUSE
RUSERUSERUSERUSERUSER
SERUSERUSERUSERUSERUSE
RUSERUSERUSERUSERUSER
SERUSERUSERUSERUSERUSE
RUSERUSERUSERUSERUSER
SERUSERUSERUSERUSERUSE
RUSERUSERUSERUSERUSER
SERUSERUSERUSERUSERUSE
RUSERUSERUSERUSERUSER